Frontiers in South and Southeast Asian Development Research

This book series publishes monographs and edited volumes devoted to studies on various aspects of development and policies in South and Southeast Asia. The books cover in-depth analyses of individual countries, regions, sectors, and comparative cross-country studies. They include both a specific and a general focus on the latest advances of the various aspects of development. In cross-country studies, the geographical scope will be inclusive to the developed Asian economies when it comes to issues like security, economic cooperation, energy, finances, trade, environment, and technology transfer. The books in this series are expected to go beyond the traditional concepts of inequality, poverty, growth, and development and offer a broad, in-depth analysis and inclusive view on sustainable development and well-being. The series provides a platform for researchers to carry out rigorous analyses, to promote, share, and discuss issues, findings, and perspectives in various areas. The topics may include, but not limited to agriculture, industry, services, business, governance, institutions, politics, security, globalization, international relations, gender, culture, religion and society, economics, and social development strategies of South and Southeast Asia. Not only that, given the current pandemic and its multifaceted impacts on trade, production, supply chain and organization of economic activities, this series also purports to highlight changes in research on development economics in the post-pandemic timeframe. Frontiers in South and Southeast Asian Development Research allows for a deeper appreciation of the various issues around Asian development with high quality and peer reviewed contributions. Volumes published in the series are important reading offering original research supported empirically which can serve as reference sources for researchers, educators, and graduate students, as well as consultants, business professionals, and policy makers, interested in the private and public sectors development of the Asian continent and its challenges and strategies for sustainable development.

Almas Heshmati • Pham Khanh Nam
Editors

Circular Economy and its Implementations in Southeast Asia

Springer

Editors
Almas Heshmati
College of Economics, Law and Government
University of Economics Ho Chi Minh City
Ho Chi Minh City, Vietnam

Pham Khanh Nam
Environmental Economics
University of Economics Ho Chi Minh City
Ho Chi Minh City, Vietnam

ISSN 2948-1279 ISSN 2948-1287 (electronic)
Frontiers in South and Southeast Asian Development Research
ISBN 978-981-96-0538-5 ISBN 978-981-96-0539-2 (eBook)
https://doi.org/10.1007/978-981-96-0539-2

This Springer imprint is published by the registered company Springer Nature Singapore Pte Ltd.
The registered company address is: 152 Beach Road, #21-01/04 Gateway East, Singapore 189721, Singapore

Contents

Editors and Contributors

About the Editors

Almas Heshmati is Research Fellow at the UEH University, Ho Chi Minh City and Emeritus Professor of Economics at Jönköping University, Sweden. He held Professor of Economics positions at the Sogang University, Korea University, Seoul National University, University of Kurdistan Hawler, and the MTT Agrifood Research (Finland). He was Research Fellow at the World Institute for Development Economics Research (WIDER), The United Nations University during 2001–2004. From 1998 until 2001, he was an Associate Professor of Economics at the Stockholm School of Economics. He has a Ph.D. degree from the University of Gothenburg (1994), where he held a Senior Researcher position until 1998. His research interests include applied microeconomics, globalization, development economics, efficiency, productivity, and growth with application to agriculture, manufacturing and services, environmental economics, energy economics, and well-being. In addition to many scientific journal articles, he has published books on EU Lisbon Process, Global Inequality, East Asian Manufacturing, Chinese Economy, Technology Transfer, Information Technology, Water Resources, Landmines, Power Generation, Renewable Energy, Development Economics, World Values, Poverty, Wellbeing, and Economic Growth.
College of Economics, Law and Government, University of Economics
Ho Chi Minh City, Ho Chi Minh City, Vietnam

Pham Khanh Nam is a senior research fellow and director of both the Environment for Development Center (EfD-Vietnam) and Economy and Environment Program for Southeast Asia (EEPSEA) at the University of Economics Ho Chi Minh city (UEH University). He also holds a leadership position in the College of Economics, Law and Government at UEH University. He received his PhD in Economics from the University of Gothenburg, Sweden. His areas of research interest are environmental and natural resources economics, climate change and experimental economics. Currently, his research focuses on forest devolution policy, marine plastic pollution, and renewable energy in Vietnam. He has published in a number of

peer-reviewed journals, including Journal of Human Resources, Journal of Public Economics, Journal of Economic Behavior and Organization, Journal of Economic Psychology, Oxford Economic Paper, Land Use Policy, Journal of Cleaner Production or Journal of Development Studies. He is a member of several task forces under Vietnam's Central Economic Committee and the Ministry of Natural Resources and Environment. He has consulted the World Bank, Asian Development Bank, IUCN, and Vietnam Environment Administration.
UEH University, Ho Chi Minh City, Vietnam

Contributors

A. A. Azlina Faculty of Business, Economics and Social Development, Universiti Malaysia Terengganu, Kuala Terengganu, Malaysia

Tuan Minh Cao School of Economics, Can Tho University, Can Tho City, Vietnam

Nguyen Ngoc Phuong Dinh University of Economics, Ho Chi Minh City, Vietnam

Nguyen Phuong Dung The Business School, RMIT University Vietnam, Ho Chi Minh City, Vietnam
LL.M International Business Law, Newcastle Law School, Newcastle University, Newcastle Upon Tyne, UK

Dang Thi Ha University of Law, Hue University, Hue, Vietnam

Timber Haaker Saxion University of Applied Sciences, Enschede, The Netherlands

Pham Hoang Ha The National Institute for Finance, Ministry of Finance, Hanoi, Vietnam

Tran Thi Dieu Ha University of Law, Hue University, Hue, Vietnam

Vu Truc Hanh The School of Law, University of Economics Ho Chi Minh City (UEH), Ho Chi Minh, Vietnam
LL.M, Master of Laws in International and European Law: International and National Legal Orders, Ghent University, Ghent, Belgium
LL.M Globalisation and Law, Faculty of Law, Maastricht University, Maastricht, The Netherlands

Pham Thi Thu Hong The National Institute for Finance, Ministry of Finance, Hanoi, Vietnam

Prajyot P. Gaonkar Goa University, Taleigao Plateau, Goa, India

Awang Noor Abd Ghani Malaysia Environmental Economic Association (MEEA), Kuala Lumpur, Selangor, Malaysia

Mahirah Kamaludin Faculty of Business, Economics and Social Development, Universiti Malaysia Terengganu, Kuala Terengganu, Malaysia

Huynh Phan Nhu Ngoc University of Economics, Ho Chi Minh City, Vietnam

Thanh Cong Nguyen National Economics University, Hanoi, Vietnam

Tu Duong Quach School of Economics, Can Tho University, Can Tho City, Vietnam

Zaiton Samdin Institute of Tropical Forestry and Forest Products, Universiti Putra Malaysia, Seri Kembangan, Selangor, Malaysia

Ha Le Thuy University of Law, Hue University, Hue, Vietnam

Ngoan Võ University of Economics Ho Chi Minh City, Vinh Long City, Vietnam

Circular Economy and Its Implementations in the Southeast Asia

Almas Heshmati and Pham Khanh Nam

Abstract This edited volume consists of a collection of contributions to the circular economy (CE) and its implementations in Southeast Asia. The focus is on waste management and recycling, regulatory, monetary, and financial policies and CE transition and its implications for sustainable development. It provides a comprehensive overview of the application of CE from various perspectives. The volume links environmental theory and practice and offers practitioners a deeper understanding of the topic and challenges in its implementation. It looks at the macro, meso, and micro context of the circular economy implementations, from the environmental, societal, and government views across countries in Southeast Asia. It presents approaches and practical examples of how to shape environmental policy to successfully implement a CE locally, regionally, and nationally. It provides analysis of deficiencies of current environmental policies and regulations and suggests integrated environmental policies for a circular economy supported by empirical evidence at the regional level. Thus, the manuscript is a timely and practical guide for sustainability researchers and policymakers in Asia in their progress to achieve circular economy and sustainable development goals in the fast-growing economies in Southeast Asia.

Keywords Circular economy · Sustainable development · Southeast Asia · Environmental policies and practices · Waste management and recycling · Regulatory and financial policies · Implementation of circular economy

---

A. Heshmati (✉)
College of Economics, Law and Government, University of Economics Ho Chi Minh City, Ho Chi Minh City, Vietnam
e-mail: almas.heshmati@ju.se

P. K. Nam
Environmental Economics, University of Economics Ho Chi Minh City, Ho Chi Minh City, Vietnam
e-mail: nam.pham@eepsea.net

A. Heshmati, P. K. Nam (eds.), *Circular Economy and its Implementations in Southeast Asia*, Frontiers in South and Southeast Asian Development Research,
https://doi.org/10.1007/978-981-96-0539-2_1

1 Introduction, Summary, and Policy Recommendations

1.1 Introduction and Summary

Almas Heshmati and Pham Khanh Nam

This edited volume consists of an introduction to and a summary of the contributions to the circular economy (CE) and its implementations in Southeast Asia and ten contributed chapters. The research is divided into three parts of contributed chapters on (1) waste management and recycling; (2) regulatory, monetary, and financial policies; and (3) circular economy transition and its implications for sustainable development. The applications are diverse and cover several countries in Southeast Asia including Malaysia, Vietnam, and regions of the Association of Southeast Asian Nations (ASEAN) and South Asian Association for Regional Cooperation (SAARC) in comparison with the European Union (EU)—the leader in environmental economics and sustainable development.

The circular economy and its implementations include studies of the transition from the linear economy, and it explores examples of the circular economy implementation in Southeast Asia which is a densely populated region with significant share of the world production. Unlike other books which majority of them provide narrow perspectives of wide-ranging aspects of circular economy, such as recycling, businesses models, waste management, and supply chains, this collected volume provides a comprehensive overview of the application of circular economy from various perspectives and different parts of Southeast Asia. Its unique and empirical insights into the approaches, methods, laws, and policies that enable countries to make the transformation to a circular economy show how recent research, trends, and attitudes mature to be practiced in actual policies that are oriented to achieve the common sustainable development.

Global consumption of some raw materials is currently exceeding the Earth's regenerative capacity. Circular economy offers a more sustainable model to raw material use, both depletable and renewable materials, and it provides new business opportunities and employment generation in green areas. A circular economy is rebuilding different forms of capital to reduce, reuse, and recycle material offering opportunities and solutions to recovery of nature and many problems faced by humans. This book is edited and written by editors and authors experienced in the field and with potential use by practitioners of sustainable development in Asia. It contains practical examples and case studies with real data that are used to evaluate implementation of circular economy.

The volume links environmental theory and practice and offers practitioners a deeper understanding of the topic and challenges in its implementation. It looks at the macro, meso, and micro context of the circular economy implementations, from the environmental, societal, and government views across countries. It presents approaches and practical examples of how to shape environmental policy to successfully implement a circular economy locally, regionally, and nationally. It provides analysis of deficiencies of current environmental policies and regulations and suggests integrated environmental policies for a circular economy supported by empirical evidence at the regional level. Thus, the manuscript is a timely and practical guide for sustainability researchers and policymakers in Asia in their progress to achieve a circular economy and sustainable development goals.

This book offers an up-to-date picture of the environment in the fast-growing economies in Southeast Asia. The countries in the region have invested heavily in development infrastructure and cooperate to develop in an environmentally sustainable way. This edited volume is written by professionals with good insights into the features of the region's economies and their development potential. The chapters provide an up-to-date empirical research-based picture of the recent trends in the region's economic development. The target audience are educators, undergraduate and graduate students, domestic and foreign investors, government agencies, provincial and industrial sector decision-makers, trade partners, libraries, and individual researchers, as well as professional and practitioners of environmental policy, circular economy, natural resource management, and sustainable development. This book helps to fill the gap in knowledge about the Southeast Asian economies and their environmental conditions and development potential.

1.2 Part 1: Waste Management and Recycling

This part consists of two chapters. The first chapter covers circular economy as an optimizer of waste management with a comparative analysis of the EU and ASEAN countries with development implications. The second chapter provides a review of e-waste estimation methods and models employed in SAARC and ASEAN member countries.

In recent decades, plastic waste has risen to the forefront of worldwide concerns, primarily due to its pervasive presence in both terrestrial and marine ecosystems. This alarming trend is attributed to the increasing consumption of raw materials, the problems associated with recycling and management of plastic material, and heightened pollution levels. Nguyen Phuong Dung and Vu Truc Hanh in second chapter conduct a comprehensive review of *Circular Economy: Optimizing Waste Management—A Comparative Analysis of the European Union and ASEAN with Development Implications*. The issue is important to the ASEAN member states. The ASEAN Community Vision 2025 aims at establishing "vibrant, sustainable and highly integrated economies." Achieving this goal necessitates departure from the traditional linear economy characterized by wasteful consumption and disposal. Instead, it calls for a transformative shift toward a sustainable and regenerative circular economy paradigm with core principles of waste reduction, recycling, and resource efficiency. This research attempts to offer the means to address the menace of plastic waste but also a pathway to sustaining economic growth while preserving and restoring the environment. Through the adoption of a circular economy framework, ASEAN nations will have the potential to harness the dual benefits of environmental preservation and economic development. This chapter conducts a comprehensive review and comparative analysis of the policy and regulatory frameworks governing the development and implementation of the circular economy in the European Union. It is offering valuable lessons and implications for enhancing circular economy practices in the ASEAN region.

Electronic waste (e-waste) refers to waste generated from the disposal of electrical and electronic equipment. A report titled "Global E-waste Monitor 2020" estimates total e-waste generation in the world as 53.6 million metric tons (MMt), which is expected to cross 74 MMt by 2030. Electrical and electronic equipment are of different types and kinds, some of which are reused, and some are not. In Chapter Prajyot P. Gaonkar conducts research entitled *A Systematic Review of E-Waste Estimation Methods and Models in SAARC and ASEAN Countries*. For empirical part of this study, SAARC and ASEAN member countries are considered. Preferred Reporting Items for Systematic Reviews and Meta-Analysis (PRISAM) guidelines are followed for this research. The methodological section includes three key stages which are identification, screening, and filtering the literature. The three broad levels of analysis are publication specific, method/model specific, and comparative analysis. Suggestions and future research issues are discussed.

1.3 Part 2: Regulatory, Monetary, and Financial Policy

This part consists of four chapters related to regulatory, monetary, and financial policy associated with circular economy and its implementation. The first chapter looks at the transmission mechanism of monetary policy and circular economy in Vietnam. The second chapter investigates the barriers of law regulation to a circular economy in SMSs and their integration in a national sustainable development strategy. The third chapter sheds light on international experience on financial policy to promote the process of collecting and recycling scrapped vehicles and draws some useful practicable lessons for Vietnam. The last chapter investigates electronic banking fraud prevention in Vietnam in the context of a circular economy.

In Chapter entitled *A Transmission Mechanism of Monetary Policy and Circular Economy in Vietnam*, Quach Duong Tu and Cao Minh Tuan examine the transmission mechanism of monetary policy and the circular economy in Vietnam. The period of study is from 2000 to 2020, utilizing a semi-structured vector autoregression (SVAR) model. The study identifies exogenous shocks to monetary policy that are beyond the control of the State Bank of Vietnam. The results indicate that monetary easing shocks lead to economic growth recovery after four quarters, but it results in a significant increase in the discount rate. Another finding is that the consumer price index initially shows no response, posing difficulties in monetary transmission. The analysis of exchange rate shocks also reveals substantial pressure on the domestic currency to depreciate. The monetary policy index closely associates changes in the State Bank's response with innovations in the exchange rate. This research highlights the intricate dynamics of the transmission mechanism of monetary policy shocks in Vietnam and their impact on implementation of the circular economy and its policy outcomes.

Circular economy in practice is broadly defined as a regenerative industrial economy that captures vast amounts of resources for repeated use while positively impacting the environment and natural resource capital. It has been linked to significant economic, environmental, and health benefits. The significant role of small and medium-sized enterprises (SMEs) in most economies provides a base for an empirical analysis that allows to explore the potential for implementation of the circular economy in SMEs. The implementation and barriers for the transition to circular economy are evidenced in the process to integrate in a sustainable development strategy. Chapter written by Ngoan Võ which is entitled *Barriers of Law Regulation to a Circular Economy in SMEs to Integrate in a Sustainable Development Strategy*. Based on the described scenario of challenges and opportunities, the study conducts a legal analysis and shows that companies are obliged to complying with the law system in Vietnam. This study shows that the barriers and opportunities have been mainly validated in general and increase attention and awareness to introduce actions necessary to encourage SME's development in line with circular economy principles.

Vehicles are the main source of transportation of persons and goods. Scrapped or end-of-life vehicle (ELV) disposal is expected soon to become a serious challenge for countries as the vehicle market in many countries continues to accelerate, while vehicle life expectancy is declining. This leads to the market for used vehicles and recycled parts expected to continue to expand. If ELV disposal is not handled appropriately, serious social and environmental problems will arise, and costly neglection will harm property and circular economy business opportunities. Therefore, to reduce waste to protect the environment, countries aim to encourage the reuse and recycling of ELV components by rapidly recovering ELVs. In Chapter Pham Hoang Ha and Pham Thi Thu Hong studied *International Experience on Financial Policy to Promote the Process of Collecting and Recycling Scrapped Vehicles and Lessons for Vietnam*. The country has been oriented to apply appropriate fiscal policy to support the recovery of discarded vehicles to enhance their reuse and recycle. The article studies the experiences of developed and developing countries in establishing fiscal policy mechanisms to contribute to the recovery of waste vehicles. An acceleration of the process of recovering discarded vehicles from manufacturers and users will provide important lessons for implementation of reuse, repair, and recycle principles of circular economy in Vietnam.

Banking systems in all countries are digitalized. Mobility of capital in the fast-growing international capital market is significantly higher than mobility of goods and labor. Digitalization has increased the speed of transactions significantly, and technology has been labor saving. Chapter authored by Ha Le Thuy is entitled *Electronic Banking Fraud Prevention in Vietnam in the Context of a Circular Economy*. The article investigates electronic banking fraud and clarifies fraud in the banking sector. The phenomenon is one of the risks arising in the circular economy. It highlights current methods of e-banking fraud through research and learns about legal regulations domestically and internationally. In addition, the article also identifies the risks and challenges for preventing e-banking fraud in Vietnam. It can be instrumental to finding effective electronic banking fraud prevention solutions to enhance implementation of circular economy in Vietnam.

1.4 Part 3: Circular Economy Transition and Its Implications for Sustainable Development

This part consists of four chapters presenting research on transition from a linear economy to circular economy. The first chapter elaborates on closing the loop by investigating how energy use efficiency labeling supports circular economy goals. The second chapter looks at the impacts of climate change on circular economy development and ensuring human rights and well-being of the population in Vietnam. The third chapter reviews the EU's legal framework for secondary raw material use, and it provides recommendations for Vietnam secondary material management. The final chapter explores the ambitions and needs of SMEs in transition to circular economy in Vietnam.

Energy labels serve as a crucial tool for educating consumers on the energy efficiency of household appliances. These energy labels enable consumers to make wise decisions that save energy and money and reduce their carbon footprint by providing precise information on appliances energy consumption. Energy-efficient appliances promote sustainability by reducing energy consumption and greenhouse gas emissions and conserving depletable and scarce resources. By conserving natural resources and minimizing resource extraction and waste generation, energy efficiency together with circular economy promotes sustainable development for future generations. Mahirah Kamaludin, A. A. Azlina, Awang Noor Abd Ghani, and Zaiton Samdin contributed Chapter entitled *Closing the Loop: How Energy Efficiency Labeling Supports Circular Economy Goals*. This study observes the Malaysian state of public awareness and knowledge of energy-efficient labeling and its acceptance. The study has collected primary data from 791 participants in selected states of Malaysia. About 75% of the samples are willing to pay more for energy-efficient appliances to support sustainability. The study found that age, employment status, and environmental awareness are important factors that drive the public willingness to spend more on energy labeled appliances. Energy labels provide information about energy consumption and the environmental impact of the products, thus supporting application of the circular economy. Recognition of the labels relevance by policymakers, industry stakeholders, and the public is crucial for employment of energy-efficient labeling. Besides, the energy labels play a vital role in accomplishing sustainable development goals 7 (SDG 7), namely, "Affordable and Clean Energy."

Climate change is said to be one of the biggest challenges humanities are facing in the twenty-first century, and Vietnam is one of the most vulnerable countries in the world. Weather phenomena such as droughts, floods, landslides, and storms have complicated developments. The risk of forest fires in the North Central and Central Highlands ecological regions is increasing. The literature shows that the risk of landslides in mountainous areas is affecting the agricultural and industrial economy, impacting on human rights and social justice. The environmental and health consequences threaten the assurance of human rights standards in Vietnam. Chapter written by Dang Thi Ha and Tran Thi Dieu Ha is entitled *Impact of Climate Change on Circular Economy Development and Ensuring Human Rights in Vietnam*. According to some studies, it is more difficult to ensure human rights standards in developing countries, and disadvantaged groups are most vulnerable to the impacts of climate change. In addition, they also suffer serious impacts from greenhouse gas emissions when living in urban areas or near industrial zones. In recognizing the seriousness of climate change, Vietnam is searching for solutions to deal with this situation. The potential measures include transition of the agricultural economy that is dependent on resources and heavy industry to a circular economy and reduction of greenhouse gas emissions which are strategic goals. The focus here is on the nexus between minimizing the impact of climate change, maintaining economic growth, and ensuring human rights standards.

Secondary raw materials refer to material that is not the primary products from manufacturing and other industrial sectors. The significance of secondary material in production is important for material use, application of circular economy, and achievement of sustainable development. Huynh Phan Nhu Ngoc and Nguyen Ngoc Phuong Dinh in Chapter focus on the issue of *EU's Legal Framework for Secondary Raw Materials and Recommendations for Vietnam: Toward Sustainable Resource Management and Circular Economy*. This study explores the European Union's legal framework for secondary raw materials and offers customized recommendations to empower Vietnam in the pursuit of sustainable resource management and adoption of circular economy principles to reduce, reuse, and recycle materials. The paper delves into key EU directives, regulations, and policies adopted to Vietnamese conditions and emphasizes their role in optimizing secondary raw materials. A comparative analysis of Vietnam's current regulations is provided, and attempt is made to identify opportunities for alignment. Additionally, the article showcases practiced examples of sustainable approach to recycled materials. The authors present a theoretical framework for circular economy implementation and methodology for secondary raw materials. Detailed statistical data on the circular material use rate in the EU and a case study on Vietnam's initial success further enrich the content. The analysis paves the way for Vietnam's transition toward a sustainable development and circular economy.

Small and medium-sized enterprises (SMEs) are considered as the backbone of economies. SMEs drive growth, provide new employment opportunities, and open new markets. SMEs contribute significantly toward gross domestic product (GDP) and supply and anchor big retail businesses with products and services. Circular business model allows SMEs to capitalize on the economic benefits of reusing resources and minimizing waste by recycling, contributing to firms' profitability, growth, survival, and environmental sustainability. The circular economy promotes resource use efficiency, reuse of resources, repairs, and recycling. In addition, the CE approach cuts costs for businesses and consumers and reduces dependence on fragile and volatile markets for raw materials. The final Chapter written by Nguyen Cong Thanh and Timber Haaker is labeled *Exploring the Ambitions and Needs of Small and Medium-Sized Enterprises in the Transition to the CE in Vietnam*. The survey results from a sample of 69 SMEs operating in the northern, central, and southern regions of Vietnam are reported in this study. The surveyed enterprises show their ambitions in implementing the CE in the future. The key issues motivating SMEs to adopt CE business model fall into economic, financial, and market categories. The key barriers to adopting CE business models relate to technological and cultural factors. SMEs in Vietnam lack experience and knowledge of CE business models. The results show a strong need for supporting SMEs in Vietnam to gain a better understanding of the CE approach. The findings point to importance of public support playing a role to enable a more effective CE adoption and implementation in Vietnam.

2 Editor Policy Recommendations

Promoting the circular economy in Southeast Asia requires a set of policy recommendations that integrate economic growth with environmental stewardship. These recommendations, derived from the ten chapters of this volume, address the unique challenges and opportunities in the region circular economy adaptation.

One key recommendation is for ASEAN member states to adopt a circular economy (CE) framework tailored to the region's needs to address the growing plastic waste problem. Drawing on the European Union's success, countries should focus on reducing, reusing, and recycling (3R principles of CE) plastic waste. This requires strong laws that make producers responsible for waste and promote sustainable waste management. Collaboration across ASEAN, such as setting common standards in using plastics and other materials and sharing knowledge, will help to protect the environment and drive green economic growth. Involving businesses and local communities is essential to make the transition to a circular economy inclusive and sustainable.

Standardized methods for estimating electronic waste (e-waste) are essential for better management in SAARC and ASEAN countries, as suggested by Chapter "A systematic review of E-waste estimation methods and models in SAARC and ASEAN countries". Governments should implement consistent and reliable methods to ensure accurate data collection and informed policy decisions. By doing so, countries can better compare e-waste data and coordinate regional efforts to address e-waste challenges. Focusing on future e-waste estimates and using advanced statistical models will help to prepare for potential increasing amounts of electronic waste. These practices will support recycling and disposal strategies, therefore reducing negative environmental and health impacts.

To effectively address the unique challenges of the circular economy, as an example, Vietnamese policymakers can develop a more resilient and adaptive monetary policy approach. Chapter “A transmission mechanism of monetary policy and circular economy in Vietnam” implies that Vietnam’s monetary policy could benefit from allowing more flexibility in the exchange rate rather than maintaining tight control. This adjustment would enable the exchange rate to better reflect actual economic conditions, leading to greater stability. Furthermore, managing inflation by addressing underlying causes, such as supply chain issues, rather than relying solely on monetary policy, would promote more stable prices and support implementation of sustainable growth. Directing increased money supply toward production sectors like manufacturing would stimulate investment and job creation, in particular green jobs, while clear communication from the central bank about inflation targets would reduce uncertainty and build confidence among business community and consumers.

Chapter “Barriers of law regulation to a circular economy in SMEs to integrate in a sustainable development strategy” suggests that simplifying and clarifying regulations is critical to helping small and medium-sized enterprises (SMEs) in Vietnam to adopt circular economy practices. Updating laws to better support recycling, waste reduction, and sustainable material use will make it easier for SMEs to engage in material circular practices. Offering incentives such as tax benefits, easier access to funding, and support for green technology innovation activities will further encourage SMEs to prioritize sustainability. Improved collaboration between the government, industry, and educational institutions is also necessary, providing SMEs with the knowledge, training, and support needed to overcome barriers related to a lack of expertise, knowledge, or resources. Building supportive infrastructure, including better waste management systems and accessible recycling facilities, will further enable the adoption of circular economy practices.

Financial policies that incentivize the proper disposal of end-of-life vehicles (ELVs) could significantly improve recycling efforts in Vietnam. Offering financial rewards for vehicle owners who take their old vehicles to authorized recycling facilities, funded by taxes or fees from vehicle sales, would make this option more attractive than informal markets. Implementing prepaid recycling fees at the point of new vehicle purchase could also support the recycling process and raise consumer awareness about more responsible disposal. Tax breaks or fee reductions for those who recycle their old vehicles and purchase new ones would encourage the shift to cleaner, more efficient models. Enhanced vehicle registration and monitoring systems would further help to manage ELV issue effectively.

Combating electronic banking fraud in Vietnam requires strengthening security measures and raising public awareness, as Chapter “Electronic banking fraud prevention in Vietnam in the context of a circular economy” concludes. The government and banks should adopt advanced technologies to protect customer data and regularly update security systems to stay ahead of evolving fraud methods. Public awareness campaigns are crucial to educate customers about common fraud tactics, such as phishing, and provide guidance on protecting personal information. Enhancing regulations and enforcement by updating laws and imposing severe penalties on fraudsters will also be necessary. Collaboration between banks, government agencies including environmental agencies, and international organizations will improve the identification and mitigation of risks.

Chapter "Closing the loop: How energy efficiency labelling supports circular economy goals" focuses on circular economy in Malaysia and suggests that boosting energy efficiency and supporting the circular economy can be achieved by promoting the widespread use of energy efficiency labels on appliances. Public awareness campaigns should educate consumers on the cost savings and environmental benefits of energy-efficient products. Making energy labeling mandatory for a broader range of products would further ensure and utilize potential energy savings. Financial incentives, such as tax rebates or discounts, can encourage consumers to choose energy-efficient options. Collaboration between the government and manufacturers is essential to ensure that energy labels are clear, accurate, and trustworthy.

Incorporating climate adaptation strategies into economic policies is vital for addressing the impact of climate change, in this case on Vietnam's development and protecting human rights. Enforcing environmental laws, setting clear regulations on emissions and waste management, and promoting sustainable practices will support a circular economy. Educating the public, particularly in vulnerable areas, about climate change and how to take protective actions is essential. Investments in resilient infrastructure, such as flood-resistant housing and early warning systems, are crucial to minimizing damage from extreme weather events. Protecting vulnerable groups by ensuring that they have access to necessary resources, including healthcare and safe housing, is a priority.

Vietnam can benefit from adopting elements of the European Union's framework for managing secondary raw materials to support the transition to a circular economy. Strengthening legal regulations similar to the EU's Waste Management Framework Directive will help reduce waste and promote active and broad recycling. Extended producer responsibility (EPR) programs, which make producers accountable for the entire life cycle of their products, encourage eco-friendly design and reduce waste. Investment in recycling infrastructure will foster a domestic circular economy and reduce dependence on imported materials. Public awareness campaigns promoting recycling and sustainable practices are also necessary, along with fostering public-private partnerships to accelerate the adoption of circular economy initiatives.

To further support small and medium-sized enterprises (SMEs) in Vietnam as they transit to circular economy practices, Chapter "Exploring the ambitions and needs of small and medium-sized enterprises in the transition to the CE in Vietnam" suggests that increasing awareness and understanding through targeted training and informational campaigns is key. Financial support in the form of subsidies or low-interest loans can help reduce the costs of transitioning to circular business models. Promoting partnerships and collaborations by creating platforms for SMEs to connect with larger firms, research institutions, and consumer networks will drive innovation and knowledge-sharing. Improving access to technology and infrastructure, such as recycling facilities and modern technologies that support CE practices, is crucial for enabling SMEs to adopt circular economy practices, contributing to sustainable development and green economic growth in Vietnam.

Acknowledgments The editors are grateful to an anonymous reviewer, Gouranga Das, and Mantu Kumar Mahalik for their comments and suggestions on an earlier version of the book manuscript and William Achauer, for his guidance and Aashika Philomi and Anushangi Weerakoon for their support during the preparation of the manuscript. The editors thank the contributing authors for their invaluable contributions to this volume.

Part I
Waste Management and Recycling

Circular Economy: Optimizing Waste Management—A Comparative Analysis of the European Union and ASEAN with Development Implications

Nguyen Phuong Dung and Vu Truc Hanh

Abstract In contemporary times, plastic waste has risen to the forefront of worldwide environmental concerns, primarily due to its pervasive presence in both terrestrial and marine ecosystems. This alarming trend is attributed to the increasing consumption of raw materials and heightened pollution levels. The ASEAN member states, in accordance with the ASEAN Community Vision 2025, sets out the ambitious objective of establishing "vibrant, sustainable, and highly integrated economies." Achieving this goal necessitates a fundamental departure from the traditional linear economy characterized by wasteful consumption and disposal. Instead, it calls for a transformative shift toward a sustainable and regenerative circular economy (CE) paradigm. The transition to a CE, with its core principles of waste reduction, recycling, and resource efficiency, assumes unparalleled significance. It offers not only the means to address the menace of plastic waste but also a pathway to sustaining economic growth while concurrently preserving and restoring the environment. Through the adoption of a CE framework, ASEAN nations have the potential to harness the dual benefits of environmental preservation and economic prosperity, thereby contributing to a more sustainable and harmonious future. This chapter will conduct a comprehensive review and comparative analysis of the policy and regulatory frameworks governing the development and implementation of the CE in the European Union (EU) and the ASEAN region. The insights gained

N. P. Dung
The Business School, RMIT University Vietnam, Ho Chi Minh City, Vietnam

LL.M International Business Law, Newcastle Law School, Newcastle University, Newcastle Upon Tyne, UK

V. T. Hanh (✉)
The School of Law, University of Economics Ho Chi Minh City (UEH), Ho Chi Minh, Vietnam

LL.M, Master of Laws in International and European Law: International and National Legal Orders, Ghent University, Ghent, Belgium

LL.M Globalisation and Law, Faculty of Law, Maastricht University, Maastricht, The Netherlands

A. Heshmati, P. K. Nam (eds.), *Circular Economy and its Implementations in Southeast Asia*, Frontiers in South and Southeast Asian Development Research,
https://doi.org/10.1007/978-981-96-0539-2_2

from this examination will be used to offer valuable lessons and implications for enhancing CE practices in the ASEAN region.

Keywords Circular economy · Plastic waste · Waste management · European Union · ASEAN

1 Introduction

Plastic, with its distinctive attributes such as lightness, durability, waterproofness, and cost-effectiveness, plays a pivotal role in the invention and production of a wide array of goods, offering heightened convenience across various domains of human life (De Wit et al., 2021, pp. 10–11; World Economic Forum, 2020, p. 6). The plastic industry's profitability is undeniably highly profitable, demonstrated by its significant growth over the years and its annual revenue of approximately USD600–700 billion, which contributed to the world economy. Despite the substantial economic benefits it offers, the plastic industry also imposes substantial costs on both the environment and society (UNEP, 2023, p. 5; De Wit et al., 2021, pp. 10–11). In recent times, it is estimated that around 200–400 million tons of plastic waste is generated yearly (De Wit et al., 2021, pp. 10–11; UNEP; World Economic Forum, 2020, p. 6). Plastic waste has emerged as an exigent issue of such magnitude that leads to the recognition of a distinctive marine microbial habit, termed the "plastisphere" (Barrowclough & Birkbeck, 2022), highlighting the gravity of plastic pollution as being uncontrollable and unmanageable.

Plastic, despite being a key driver of economic expansion, adheres to a linear economy paradigm known as "take-make-waste." This model is associated with critical concerns including the exhaustion of natural resources, generation of waste, environmental degradation, contributions to climate change, and negative effects on human health. Balancing the ongoing production and consumption of natural resources, which leads to resource depletion, pollution, climate change, and biodiversity loss, while also ensuring environmental protection, presents a formidable task.

In response to this challenge, the CE is considered a replacement for the one-way economy and an effective approach to attain sustainable goals (Kopnina & Poldner, 2021, p. 1). In contrast to the linear economy which operates on the assumption of an unlimited supply of resources that can be extracted without restriction, converted into goods, consumed, and ultimately disposed of in landfills, the CE purports to establish a self-sustaining system where the flows of materials and energy within the economy are closed and continuously reused. At its core, in developing countries, the CE is underpinned by a tripartite model, commonly referred to as the 3R paradigm (reduction-reuse-recycle). This approach includes three pillars, in particular the reduction of materials and energy necessary for production; the extension of product lifespans through practices such as sharing, repair, and reuse; and the recycling of materials at the end of life cycle within an ongoing and environmentally responsible cycle (Aggeri, 2020, p. 3; Kopnina & Poldner, 2021, pp. 1–2; Ghosh, 2020, pp. 2–4).

The CE is not a new concept. It can be traced back to the 1970s–1990s when it emerged from the foundation cradle-to-cradle concept and the principle of circularity in material and energy flows. It hinges on the notion that humans must integrate themselves within an ecological cyclical system capable of indefinitely regenerating any material form. By the early 2010s, the concept became more prevalent within policymaking circles (Ghosh, 2020, p. 5; Aggeri, 2020, p. 4; Kopnina & Poldner, 2021, pp. 2–5). The EU is one of the pioneers shifting toward a regional CE. It has encouraged its members to adopt relevant policies to ensure alignment with well-established policies developed over the course of several years, focusing on waste management, product policies, and resource efficiency. The EU officially issued the CE Action Plan (CEAP) and Circular Policy in late 2015 and early 2016. The policy development related to the CE has progressed significantly, with the EU achieving notable success in reducing waste generation and recycling, which are integral aspects of the CE (Aggeri, 2020, pp. 4–5; Farmer, 2020, pp. 393–400).

This chapter primarily employs comparative and analytical approaches to illuminate the current landscape of plastic waste and waste management in the EU and the ASEAN region. It explores the development and implementation of CE policies in the EU, with the intent of comparing with the ASEAN region. Although ASEAN has demonstrated an aspiration to transition toward a CE and has laid preliminary groundwork, it remains in a nascent stage with ample potential for further development. By delving into the EU's trajectory of policy development and implementation of the CE as a basis for comparison with the ASEAN region, this chapter aims to provide lessons from the EU's experience that are useful for ASEAN in constructing and enhancing regional policies pertaining to the CE to facilitate the implementation of CE at both regional and member state levels.

2 Overview of Plastic Waste and Waste Management

2.1 Plastic Waste and Waste Management in the EU

The EU serves as an ideal model for the successful implementation of a CE, especially in the realm of waste management. Concrete actions within the context of CE have been initiated in the EU since 2014, although relevant aspects had been introduced in EU regulations as far back as the 1970s (Mazur-Wierzbicka, 2021). As examples, we can point to regulations related to waste, such as Directive 94/62/EC on packaging and packaging waste dated December 20, 1994; Council Directive 75/442/EEC from July 15, 1975, which addresses waste management; Regulation (EEC) No 259/93 concerning the supervision and control of waste shipments within, into, and out of the European Community; as well as Council Directive 1999/31/EC regarding landfilling of waste. These regulations serve as a clear and concrete reflection of the EU's dedication to managing waste in a highly efficient and sustainable manner.

In addition to the legislative endeavors, the EU's accomplishments can be substantiated by empirical data and quantifiable evidence. In 2019, according to the first Global Plastics Outlook by the OECD, European countries within the OECD boasted the world's highest rates of plastic waste recycling, standing at 14%, which is notably higher than the global average of 9% (OECD, 2022). The percentage of mismanaged and uncollected plastic litter in Europe is also relatively low on a global scale, at just around 5%, significantly lower than the world's average of 22% (OECD, 2022). Even non-OECD European nations exhibit commendable figures. For instance, the rate of mismanaged and uncollected plastic litter in these countries is only 11%, in stark contrast to other non-OECD Asian countries where this figure stands at 45% (OECD, 2022). This means that nearly half of the plastic waste in these Asian countries is inadequately managed and collected.

Furthermore, in terms of progress over time, the EU has demonstrated significant advancements, as evidenced by data related to packaging waste management. Between 2009 and 2019, the amount of plastic packaging waste generated per person in the EU increased by 24% (equivalent to an additional 6.7 kg per person) (Eurostat, 2021). Simultaneously, the volume of recycled plastic packaging waste saw a substantial increase, soaring by 50% (equivalent to an additional 4.7 kg per person) over the same period (Eurostat, 2021). In 2020, the average EU citizen generated nearly 35 kg of plastic packaging waste, which is equivalent to 1070 bottles, based on the average weight of a 500 ml PET bottle (Kálmán, 2023). Notably, the most optimistic estimate reveals that an impressive 39% of discarded plastic packaging is effectively recycled (Kálmán, 2023). These numerical insights illustrate the EU's commitment to improving waste management practices and its dedication to transitioning toward a CE.

Yet, some believe that this statistic should see enhancements, and, of course, challenges persist. These challenges include substantial variations between member states in their commitment to integrating EU waste law into their national framework, as well as stark disparities in the enforcement and compliance with the adopted measures (European Environmental Bureau, 2020). Nevertheless, one cannot deny the EU's pioneering efforts in promoting a CE on a global scale.

The EU's achievements can be attributed to a multitude of factors, with a prominent one being the well-established CE policy that has evolved over several years, starting from 2015 to the present day. This part aims to provide a concise overview of significant milestones within the CE in chronological order, allowing us to emphasize the EU's most significant advancements over time, with a particular focus on waste management, notably plastic waste. It will then proceed to dissect crucial principles of key legal instruments in waste management, focusing on the Waste Framework Directive (WFD), while also exploring their interconnectedness with other instruments such as landfill regulations and waste shipment. Following that are exemplary case studies from countries recognized for their successful implementation of the EU regulations.

2.2 Plastic Waste and Waste Management in ASEAN

According to statistics from 2016, apart from Indonesia, other ASEAN member states (AMS) were not the primary contributors to plastic waste production (World Population Review Plastic Pollution by Country, 2023). Despite not holding the high rank in plastic waste generation, all AMS have encountered difficulties in effectively managing plastic waste, leading to its release into the oceans. Over recent years, the plastic waste issue in ASEAN has escalated in severity, especially in the wake of China's 2017 ban on importing plastic scraps. Consequently, AMS, such as Malaysia and Vietnam, have become primary destinations for plastic waste exports from the EU, the USA, and Japan (Akenji et al., 2019; EU-ASEAN Business Council, 2020, p. 6). The plastic pollution is further compounded by rapid urbanization, with cities being accountable for more than 60% of plastic waste leakage into the environment (Akenji et al., 2019; EU-ASEAN Business Council, 2020, pp. 6–7).

The data recorded in recent years highlights a disconcerting trend observed in the majority of AMS. Effective management of plastic waste remains a formidable challenge, as evidenced by the fact that six out of the top 10 nations contributing the highest volumes of plastic waste to the oceans are AMS. Failure in plastic waste management has resulted in the discharge of substantial amounts of plastic waste into the ocean. Roughly 80% of plastic waste that enters the oceans from rivers worldwide is partially attributed to urban rivers in Southeast Asia (Meijer et al., 2021).

As per the study conducted by Jambeck et al. (2015), the primary origin of plastic waste emitted into the oceans is land-based, with ASEAN countries significantly contributing to this issue. On average, Indonesia (3.22 MMT[1]/year), the Philippines (1.88 MMT/year), Vietnam (1.83 MMT/year), Thailand (1.03 MMT/per year), Malaysia (0.94 MMT/year), and Bangladesh (0.79 MMT/year) were identified as major contributors to the mismanaged plastic waste (MPW) (Jambeck et al., 2015, p. 769). In addition, according to the most recent data by 2021, the Philippines ranked first in this regard, with a staggering 356,371 tons of MPW. Malaysia followed closely, securing the third position with 73,098 tons of MPW. Indonesia occupied the fifth place with 56,333 tons of MPW, while Vietnam held the seventh rank, accounting for 28,221 tons of MPW. Additionally, Bangladesh was placed eighth in this world ranking, with 24,640 tons of MPW, and Thailand followed closely behind, ranking ninth with 22,806 tons of MPW (World Population Review Plastic Pollution by Country, 2023).

The gravity of the MPW situation in those ASEAN countries is ascribed to waste management infrastructures which are underdeveloped and heavily rely on the informal sector operations. Furthermore, waste management systems hinge on local leadership and are vulnerable to political influences and cycles. In rural areas, the responsibility for waste management falls upon local communities. However, there are not enough financial resources, institutional capacity, and technical

[1] MMT: million metric ton.

expertise necessary for forming and operating effective waste management. Moreover, governments are not mandatorily required to furnish comprehensive waste services, and households lack incentives to engage in accountable waste management (UNEP, 2023, p. 32).

Waste mismanagement, particularly, mismanagement of plastic waste, has additionally caused economic loss to the region. By 2019, several industries in the Asia-Pacific region such as tourism, fishing, and shipping incurred annual losses totaling USD1.3 billion/year as a repercussion of waste mismanagement (EU-ASEAN Business Council, 2020, p. 6). More specifically, failure in plastic recycling also costs ASEAN nations' economies substantial monetary amounts. The World Bank's reports estimate economic losses incurred by certain AMS that are responsible for the mismanagement of plastic litter, leading to its leakage into the oceans, in particular Thailand (USD3.6–4.0 billion/year) (World Bank, 2021a), Malaysia (USD1–1.11 billion/year) (World Bank, 2021b), the Philippines (USD0.79–0.89 billion/year) (World Bank, 2021c), and Vietnam (USD2.2–2.9 billion/year) (World Bank, 2021d).

The conundrum of achieving both economic development and environmental protection prompts a critical issue that how AMS are going to strike this balance. Transitioning to a CE stands out as a potential approach. The crux of the issue is how AMS can set out a CE and harness the potential advantages it offers.

3 Overview of the Policy Development and Implementation of the CE

3.1 The CE in the EU

The year 2015 carried profound significance for CE within the EU. It saw the emergence of the CEAP, a comprehensive framework of legislative and nonlegislative measures, which was adopted to facilitate the transition from a linear to a circular economic model (Ellen MacArthur Foundation, 2022). Its overarching goals were multifaceted, aiming to boost global competitiveness, nurture sustainable economic growth, catalyze innovation, and create employment opportunities (European Commission, 2018).

Notably, the Action Plan marked a pivotal moment as the first Commission program that sought to promote a holistic approach across the entire value chain, encompassing the design and production phases, consumption, and waste management, involving various stakeholders such as businesses, consumers, national authorities, and European institutions (European Commission, 2018). To achieve these objectives, the Action Plan outlined 54 distinct actions, complemented by four legislative proposals related to waste management (United Nations, 2020).

One crucial legislative advancement was the issuance of the Plastic Bags Directive, which served as an amendment to the Packaging and Packaging Waste

Directive (94/62/EC) and was explicitly crafted to confront the issue of unsustainable consumption and utilization of thin plastic carrier bags (European Parliament, European Council, 2015a). These bags, characterized by their thickness falling below 50 microns, had gained notoriety as one of the top 10 frequently littered items in the EU (European Commission). As such, the Directive imposed an obligation on member states (MS) to implement a range of measures, which could encompass the potential establishment of national reduction targets, the introduction of economic tools like fees and taxes, and even the enforcement of marketing restrictions, including bans. It is worth emphasizing that, relative to other domains, legislative proposals for altering EU waste management had advanced considerably during this stage, having undergone thorough consideration and preparation for implementation (de Römph, 2018).

As part of the initial CEAP, other significant revisions to waste legislation and strategy took place in 2018. An integral part of this plan was the adoption of the European Strategy for Plastics in a Circular Economy, which formed a critical component of the broader CEAP within the EU (European Commission, 2018). This strategy is further built upon existing measures designed to curtail plastic waste, which sets out some actions, such as the adoption of the Single-Use Plastic Directive (EU) 2019/904, which primarily focuses on mitigating the environmental and human health impacts of specific plastic products (European Parliament, European Council, 2019).

Significantly, this directive sought to address the pervasive issue of plastic pollution in the aquatic environment, given that more than 80% of marine litter comprises plastics (European Commission, 2021a). The directive also targeted the ten most common single-use plastic items polluting EU beaches, accounting for the majority of marine litter in the region (Oberpriller, 2021). This underlines the EU's determination to combat plastic pollution, echoing President von der Leyen's warning that "by 2050, there will be more plastic in the ocean than fish. We need to get serious about how we turn the tide" (European Commission, 2020a).

By July 2021, EU MS were expected to prohibit the sale of several single-use plastic items and introduce labeling requirements for certain remaining products, accompanied by annual awareness campaigns (European Parliament, European Council, 2019). Notably, single-use plastic plates, cutlery, straws, balloon sticks, cotton buds, cups, food, and beverage containers made of expanded polystyrene and oxo-degradable plastic products were barred from the markets of EU MS starting from July 3, 2021 (European Commission, 2021a). Additionally, this measure aimed to achieve quantifiable reductions in the consumption of single-use plastic cups and food containers by 2026, compared to 2022, and necessitated the establishment of extended producer responsibility schemes for various single-use plastic categories beginning in 2024 (European Commission, 2021a).

In 2019, the European Green Deal was introduced, envisioning a Circular Economy Action Plan 2.0 that encompasses a sustainable product initiative (European Environmental Bureau, 2019). This strategic framework prioritizes waste prevention and envisions a CE that prevents waste and maximizes the use of resources within the EU economy (European Environmental Bureau, 2019). Part of this effort involves revising EU legislation on packaging and packaging waste, which includes mandatory targets for increased use of recycled plastics in packaging (Directorate-General for Environment, 2022a). These targets will require producers to incorporate specified levels of recycled content in new plastic packaging, transforming recycled plastic into a valuable raw material, as exemplified by PET bottles under the Single-Use Plastics Directive (European Commission, 2022).

In March 2020, the Circular Economy Action Plan 2.0 was officially launched, which stands as a cornerstone of the European Green Deal, which represents Europe's new roadmap for sustainable growth (United Nations, 2020). The updated action plan introduces a range of initiatives encompassing the entire life cycle of products, emphasizing product design, promoting CE practices, encouraging sustainable consumption, and prioritizing waste prevention to retain resources within the EU economy as long as possible (European Commission, 2022).

In 2023, the Commission recently revisited the CE monitoring framework, which was originally adopted in 2018. This revisitation showcases a dedication to refining and enhancing the tools used to measure and promote circular practices. Specifically, the Commission implements measures under EU chemical legislation REACH to restrict intentionally added microplastics in products, effectively preventing the release of approximately half a million tons of microplastics and prohibiting the sale of both raw microplastics and products intentionally infused with microplastics that release them during use (European Commission, 2023). Additionally, the creation of a brochure detailing EU actions against microplastic pollution serves as a resource to increase awareness and facilitate necessary changes (Directorate-General for Environment, 2023).

Given its intricate and extensive legal structure, finding an ideal categorization for the European legal framework in the context of waste management can be challenging. However, it becomes evident that there may be four primary categories related to waste management. These categories encompass (1) the framework of waste legislation, which includes pivotal regulations like the WFD and Waste Shipment Regulation; (2) the criteria for determining the end of waste, with the anticipated completion of technical assessments for plastic waste expected by 2024; (3) regulations governing waste management operations, such as the Landfill Directive; and (4) regulations specific to distinct waste streams, exemplified by the Directive on Packaging and Packing Waste (Antico, 2020).

Within this classification, the Waste Framework Directive (WFD) stands out as a crucial element that significantly contributes to improving the overall environmental aspects of waste management, in full accordance with the waste hierarchy (Directorate-General for Environment, 2022b). According to Article 4 of the WFD, the waste hierarchy places the highest priority on waste prevention and reuse as the most effective means of enhancing resource efficiency and reducing environmental waste impact. Ideally, waste should be prevented, and what cannot be prevented should be maximally reused, recycled, and recovered, with disposal, especially landfilling, kept to a minimum (Institute for European Environmental Policy, 2010). This principle is particularly significant when considering the environmental impact of packaging waste. It is both a major consumer of virgin materials, with 40% of plastics in the EU dedicated to packaging, and a primary contributor to waste, constituting 36% of municipal solid waste (The European Files, 2022).

In response to this, the Packaging and Packaging Waste Directive reinforces these core ideas, specifying actions in Article 4 to diminish plastic packaging generation and in Article 5 to encourage the adoption of reusable packaging in the market (European Parliament, European Council, 2015b). The importance of this principle is noteworthy, especially when looking at our counterparts in Southeast Asia, where waste mismanagement is prevalent. Perhaps it is a valuable notion to start by focusing on waste prevention at the source.

Another vital principle is the preservation of both public health and the environment, as articulated in Article 13 of the WFD. This article stipulates that all waste must be managed in a manner that is environmentally responsible and safe. Furthermore, it places the responsibility on EU MS to implement necessary measures to ensure that waste management practices do not jeopardize human health or harm the natural world. A striking illustration of the practical application of this principle can be seen in the case of illegal landfills in Italy, which carries substantial financial consequences.

It is worth noting that this legal action is not grounded in the Landfill Directive but rather finds its foundation in Article 13 of the WFD (Antico, 2020). The European Court of Justice, in its ruling of October 15, 2014, underscored the importance of selecting the most suitable waste treatment methods before landfilling, with the aim of minimizing adverse impacts on both the environment and human health (EU Monitor, 2021). The imperative of safeguarding health and the environment extends beyond landfills and spreads to the domain of waste shipment, which is also subject to stringent regulations under this very principle.

The EU has implemented a robust set of rules governing the international shipment of plastic waste, aligning itself with the decisions reached by 187 countries during the 14th Conference of the Parties of the Basel Convention in May 2019 (European Commission, 2021b). The Delegated Regulation (EU) 2020/2174 amending the Waste Shipments Regulation, adopted in 2021, includes a ban on exporting plastic waste from the EU to non-OECD countries, with the exception of clean plastic waste intended for recycling (European Commission, 2020b). These regulations hold particular significance due to the rising trend of waste export to various Southeast Asian countries, especially following China's ban on waste imports in 2018 (Jain, 2020). These nations often lack the capacity and standards for sustainable waste management. Plastic waste is consequently at risk of becoming marine litter in Southeast Asian countries (Whitehead, 2018).

Therefore, by prohibiting the export of plastic waste, except for clean plastic waste sent for recycling to non-OECD countries, the EU is showing its dedication to responsible global waste management practices, ultimately aimed at mitigating the adverse environmental and human health effects of waste. These measures may even see further development, as there is a growing demand for a complete ban on plastic waste exports, primarily due to the fact that a significant portion of Europe's plastic waste intended for recycling, particularly polyethylene, often undergoes no recycling process at all (Dunphy, 2020).

The EU's success is not confined to mere words and legislation, but rather, it manifests in the real-world performance of its MS. For example, Germany has earned worldwide recognition as a recycling leader. In a pioneering move dating back to 1991, it laid the foundation for producer responsibility by implementing packaging waste regulations, making it the first EU MS to do so (Fischer, 2013). This fundamental principle, enshrined in German waste legislation, dictates that producers bear the responsibility for their products when they become waste (Fischer, 2013). Germany's journey toward becoming a recycling leader did not stop there. Regarding legal measures, to align with the waste hierarchy's five-step approach, the newly enacted Commercial Wastes Ordinance (GewAbfV), effective since August 1, 2017, establishes a strict set of obligations prioritizing the separate collection and recycling of various materials, including plastics (Federal Ministry for the Environment, Nature Conservation, Nuclear Safety, and Consumer Protection (BMUV), 2023).

The Packaging Act was also revised in 2021, which aimed to encourage the prevention of packaging waste, especially within the context of a well-functioning CE, and entailed prohibiting the sale of lightweight plastic carrier bags starting from January 1, 2022, to reduce unnecessary packaging waste (Library of Congress, 2022). Additionally, as of January 1, 2023, eateries offering takeaway food and beverages must adopt reusable packaging, with the aim of replacing single-use plastic packaging (CMS Law, 2023). Moreover, recycling rates in Germany have consistently been impressive for many years.

By 2002, the country was already recycling a remarkable 56% of its waste, and this percentage has continued to climb since then (Okafor, 2023). Germany currently holds the top spot in recycling municipal waste within the EU. For instance, in 2008, under the original WFD, the target for all MS was to achieve a minimum recycling rate of 50% for certain materials by 2020 (European Parliament, European Council, 2008). The German federal government, in 2012, raised the bar further by enacting the Circular Economy Act, imposing a higher 65% recycling rate (Igini, 2022). Subsequently, the amended WFD in 2018 established progressive recycling targets for municipal waste across Europe: 55% by 2025, 60% by 2030, and 65% by 2035 (European Parliament, European Council, 2018). Remarkably, Germany had already surpassed the latter goal well in advance, with its estimated recycling rate reaching approximately 67% in 2019 (European Environment Agency, 2022). In 2020, an astounding 99.7% of plastic packaging waste in Germany was effectively recovered, marking a remarkable increase from a mere 11.6% in 1991 (Federal Ministry for the Environment, Nature Conservation, Nuclear Safety, and Consumer Protection (BMUV), 2023). This swift transition, achieved in less than two decades, demonstrates a remarkable and praiseworthy achievement.

3.2 The CE in the ASEAN Region

Initiating the implementation of a CE can find a promising foothold in the management of plastic debris. Within the CE paradigm, plastics are purposefully engineered to have durability, repairability, resource-sharing, and efficient collection and recycling at every phase, which can lay the groundwork for other sectors in the economy (Witoonchart & Thampanishvong, 2023, p. 20). The urgent challenges related to marine litter and plastic pollution have acted as a catalyst, expediting the integration of CE principles within the ASEAN region (Bengtsson & Schroder, 2021, p. 16).

Considering the profound issues of plastic pollution and waste management within the ASEAN region, it is imperative for ASEAN to transition toward a CE as a strategic avenue to attain the sustainable objective enshrined in the ASEAN Community Vision 2025, which envisions "vibrant, sustainable, and highly integrated economies" (ASEAN, 2025, para. 5). Aligned with this vision, the ASEAN has devised and embraced the Framework for Circular Economy for the ASEAN Economic Community (ASEAN CE Framework) at the 20th AEC Council Meeting held on October 18, 2021. This ASEAN CE Framework is built upon five strategic pillars, in particular (1) harmonization of standards and the mutual recognition of circular products and services; (2) trade openness and trade facilitation in circular goods and services; (3) enhanced role of innovation, digitalization, and emerging/green technologies; (4) encouragement of competitive sustainable finance and the stimulation of innovation ESG (environmental, social, and governance) investments; and (5) efficient utilization of energy and other resources (ASEAN, 2021a).

The adoption of this framework underscores ASEAN's steadfast commitment to the exploration of new prospects and the establishment of collaborations that encompass various ASEAN sectors, dialog partners, and private entities, with the overarching goal of advancing and hastening the region's shift toward a low-carbon economy (ASEAN, 2021a, 2021b; Bengtsson & Schroder, 2021, pp. 17–18). The recent launch of the ASEAN Circular Economy Stakeholder Platform (ASEAN CE Platform) in June 2023 marks a noteworthy milestone in the partnership between ASEAN and the EU. This platform's inception holds the potential to facilitate dialogs and foster collaborative activities between the EU and ASEAN, with a key focus on advancing sustainable growth in ASEAN through the development and implementation of a CE. It provides a valuable conduit for ASEAN to gain insights and exchange experiences with the EU, which has already made substantial progress in its transition toward a CE (ACESP, 2024; EU, 2023).

Other regional instruments aimed at addressing marine plastic waste that ASEAN has endorsed recently between 2019 and 2021 are the Bangkok Declaration on Combating Marine Debris (ASEAN, 2019a), the ASEAN Framework of Action on Marine Debris (ASEAN, 2019b), and the ASEAN Regional Action Plan for Combating Marine Debris (RAP) (ASEAN Secretariat, 2021) (World Bank, 2021d, p. 46). These regional instruments targeted at marine plastic pollution reflect a strong political dedication and advocate a holistic approach firmly grounded in the CE principles. In particular, the RAP outlines 14 specific actions distributed across 3 critical phases of the value chain. These encompass strategies to curtail the introduction of plastics into the system, enhance collection processes while minimizing leakage, and create value through the reuse of waste (ASEAN Secretariat, 2021; Bengtsson & Schroder, 2021, p. 18).

At the national level, AMS emphasize the promotion of economic growth that aligns with environmental sustainability. Several AMS, such as Thailand, Malaysia, Indonesia, Vietnam, Cambodia, and Laos PDR have placed the CE at the core of their strategies for socioeconomic development, green economy, green growth, plastic waste management, and overall green and sustainable development. The CE has been formally and explicitly integrated into the national policies of certain AMS. For example, Cambodia has initiated a National Circular Economy Strategy and Action Plan in addition to the establishment of a platform for private sector engagement. Thailand has embedded the CE concept within its Bio-Circular-Green (BCG) Economy initiative, positioning the CE as an innovative catalyst for sustainable economic growth. Vietnam has issued its new Law on Environmental Protection in 2020 explicitly designating the CE as a pivotal policy tool and a foremost approach (Witoonchart & Thampanishvong, 2023, p. 20; Bengtsson et al., 2021, pp. 43–44; Bengtsson & Schroder, 2021, p. 16; Akenji et al., 2019).

Thailand stands as a leading pioneer in ASEAN for directly articulating the CE in its national policy—"BCG in Action: Circular Economy (2020)." This strategic plan has identified specific subsectors with pronounced circularity potential, including plastics, food and agriculture, and building materials. The BCG Economy has been subsequently introduced in alignment with the Bangkok Goals of the Asia-Pacific Cooperation (APEC) in 2022, emphasizing four key sectors with substantial economic promise, including agriculture and food; energy, material, and biochemicals; wellness and medicine; and tourism and the creative economy. Under this BCG Economy scheme, Thailand strives to achieve a complete recycling rate of 100% for PET and PE plastics by 2030 through a series of strategic measures, including the allocation of financial resources to create "waste hubs," the provision of support to emerging recycling business, extended producer responsibility (EPR) programs, and regulatory reform pertaining to the use of recycled plastics in food packaging (Witoonchart & Thampanishvong, 2023).

Other countries such as Indonesia, Malaysia, the Philippines, Thailand, and Vietnam have additionally set out their national plans in response to plastic pollution and waste management. In particular, Indonesia introduced "Radically Reducing Plastic Pollution in Indonesia: A Multi-Stakeholder Action Plan" in 2021, presenting a clear path to reduce plastic waste entering Indonesia's oceans by 70% by 2025 and a long-term objective to achieve minimal plastic pollution by 2040 by shifting toward a CE for plastic waste management. To guide resource and waste management and recognize marine plastic pollution, the Philippines has issued the "National Plan of Action for the Prevention, Reduction, and Management of Marine Litter" with the aim of completely eliminating waste in the Philippine waters by 2040. Malaysia and Thailand have launched their respective national roadmaps on zero single-use plastics and plastic waste management for the 2018–2030 period. The 3R model is applied through both mandatory and voluntary initiatives, promoting the active involvement of local communities and entailing coordinated actions between the private and public sectors (Bengtsson & Schroder, 2021, p. 16; Bengtsson et al., 2021, pp. 43–44; Tangwanichagapong et al., 2020, pp. 366–367).

Since 2009, Vietnam has laid out the "National Strategy for Integrated Solid Waste Management to 2025, with a Vision for 2050" as a blueprint for waste management underpinned by the 3R model and aims to drastically reduce the reliance on landfills by the year 2050. To combat the problem of marine plastic pollution, Vietnam has taken a proactive step by adopting the "National Action Plan for Management of Marine Plastic Litter by 2030." This plan outlines specific goals, including the reduction of plastic waste in Vietnamese rivers, waters, and oceans resulting from fishing, tourism, and other land-based activities by as much as 75% by 2030 (RKC-MPD; Akenji et al., 2019, p. 25; Vu et al., 2021, pp. 138–139).

As some AMS are in the process of crafting and complementing policy frameworks to shape and guide socioeconomic progress toward greater sustainability and resilience, they can draw upon the experiences and lessons from other nations. In ASEAN, due to discrepancies in each nation's economic composition, trade dynamics, and other national attributes, each AMS customizes the CE model to be suitable for specific priorities and needs (Bengtsson et al., 2021, p. 41).

ASEAN countries have formulated and adopted policies with explicit intentions to promote the CE. Some of these initiatives have comprehensive goals spanning across various economic sectors, while others are more focused on specific sectors or material categories. These policies encompass various approaches, such as EPR initiatives, recycling mandates, and sustainable product policies. Nevertheless, these policies have been operated in isolation, without integration into comprehensive policy frameworks or strategic roadmaps. Certain AMS concentrate on plastic-related concerns to formulate national strategies and action plans which prominently adhere to the CE principles (Bengtsson & Schroder, 2021, p. 16; Bengtsson et al., 2021, p. 45).

Several AMS including Indonesia, Malaysia, the Philippines, Singapore, Thailand, Vietnam, and others have considered and are in the process of deliberating and incorporating EPR for plastic packaging in their domestic legislation and policies. EPR has been enshrined in Vietnam's new Law on Environmental Protection 2020 (National Assembly of Vietnam, 2020, art. 54) and the guiding Decree 08/2022/ND-CP (Government of Vietnam, 2022) imposing the responsibility for collection and recycling of used plastic packaging on corporations.

Additionally, there will be taxes imposed on plastic bags for companies, and fines will be levied for failure to properly classify waste. Singapore has passed the Zero Waste Master Plan, in conjunction with the Resource Sustainability Act (RSA) 2020, requiring producers to bear the costs associated with the collection and treatment of their products. While discussions regarding EPR have taken place in several countries, the adoption of EPR-based recycling policies remains in its nascent stage, with the paradigm being relatively pristine in the policy landscape and not yet implemented in practice. This underscores the need for the development of a more sophisticated and all-encompassing framework to facilitate the implementation of EPR (Akenji et al., 2019, pp. 23–25; Bengtsson et al., 2021, p. 43).

4 Lessons and Implications for the Development of CE in the ASEAN Region

In contemporary times, it is evident that ASEAN has recently made some progress in formulating policies for transitioning to a CE in the region. However, while there have been significant steps toward the formulation of policies on CE, the implementation of such regional strategies remains fragmented and inconsistent. The fragmentation and absence of policy uniformity in the ASEAN can be ascribed to various underlying factors. It is important to acknowledge that ASEAN is different from the EU in terms of the nature of its association among AMS. ASEAN has not pursued deep economic or political integration. According to the ASEAN's Charter, ASEAN operates in line with the core principles of "respect for the independence, sovereignty,. . ., territorial integrity and national identity" (ASEAN, 1967, art. 2. (a)) and "non-interference in the internal affairs" (ASEAN, 1967, art. 2. (e)) of AMS. In addition to the diversity of cultures and ethnicities, differences in political governance, and varying socioeconomic situations among AMS, those aforementioned norms constitute the "ASEAN Way." This approach is characterized by a consensus-driven decision-making process and a strong emphasis on cooperative programs. The disparate political and economic backgrounds of AMS may lead to varying interpretations of the "ASEAN Way." Furthermore, ASEAN underlines individual national implementation of policies and initiatives, rather than hinging on a centralized and authoritative regional agency or bureaucracy (Koh & Robison, 2002, p. 676; Koh & Bhullar, 2011, p. 85; Boer, 2016, pp. 116–117; Mahaseth & Subramaniam, 2021, pp. 116–117).

In addition, throughout its history of development, almost all instruments adopted within ASEAN have been in the form of soft law documents. The consensus-driven approach within ASEAN results in the adoption of instruments being categorized as "soft law," indicating that they are nonlegally binding. Environment-related management such as plastic waste management was not a primary focus during ASEAN's initial establishment. This stems from the early emphasis on fostering economic, social, and cultural cooperation founded upon principles of equity and partnership, obviating the necessity for a formal treaty system. ASEAN has prioritized programmatic cooperation over the adoption of formal and legally binding instruments (Koh & Robison, 2002, p. 643; Mahaseth & Subramaniam, 2021, pp. 105–106).

The predominance of nonlegally binding instruments within ASEAN gives rise to a challenge regarding the effectiveness of enforcement mechanisms. These instruments lack the imposition of legal duties, instead placing an expectation on governments to adhere to them. There additionally exist diverse understandings and interpretations among AMS. The inclination toward the adoption of nonlegally binding instruments within the ASEAN framework serves as an additional manifestation of the "ASEAN Way," grounded in the principles of sovereignty and noninterference (Koh & Robison, 2002, pp. 642–644, 665, 677–678; Boer, 2016, pp. 116–117; Maggio, 2019, p. 209; Mahaseth & Subramaniam, 2021, pp. 115–117).

From the review of recent literature pertaining to CE and plastic waste management in ASEAN, a key observation is highlighted that each AMS has independently instituted varying policies and regulations in addressing the front challenge of plastic waste, resulting in lack of consistency and a significant disparity in the extent to which these policies have been put into practice. Certain governments such as Indonesia, Malaysia, Thailand, Myanmar, and Vietnam have devised national strategies or plans dealing with plastic waste. These strategies are considered guidelines for relevant government agencies and stakeholders to facilitate coordinative and concerted actions (Akenji et al., 2019, pp. 23–25).

However, some other AMS have yet to formulate or are still in the process of formulating these strategies like Brunei, Cambodia, the Philippines, Laos PDR, and Singapore. Single-use plastics are not yet comprehensively prohibited in most of AMS. In cases where bans or charges on disposable plastic products have been introduced, they are often restricted to items like shopping bags and are limited in effect to specific cities or municipal regions (Akenji et al., 2019, pp. 23–25). Thus, it would be challenging for ASEAN to promote and adopt a CE consistently at both the regional and national levels.

The introduction of the CE concept within ASEAN is still in its infancy, having been recently unveiled in October 2022. CE policies and regulations are in the initial phases and require further development. It is important to note that the ASEAN Circular Economy Framework characterized as "soft" law and the "ASEAN Way" present particular challenges to the transition to the CE in the region, notably the absence of robust enforcement and monitoring mechanisms. In contrast, the EU has established legally binding regulations to support the CE. For instance, EPR policies play a central role in waste management and material utilization across various product categories within the EU, including packaging, vehicles, and electronic goods. The EU legislation ensures the implementation of EPR policies by requiring EU MS to establish national-level EPR schemes. Specifically, in the context of material circularity in vehicles, the EU introduced the End-of-Life Vehicles (ELV) Directive in 2000 (Directive, 2000/53/EC), which mandates EU MS to achieve specific targets for vehicle reuse and recycling.

EU MS were obligated to enact new legislation, compel manufacturers to take back old vehicles, and develop the necessary infrastructure to meet the ambitious targets outlined in the ELV Directive. In managing packaging waste, the EU also issued Directive 94/62/EC in 1994 (Farmer, 2020, pp. 401 and 403). EU's success in developing the CE model is not solely attributed to its legal framework but also owes a significant part of its achievements to the compliance assurance system. Under Article 17 of the Treaty on European Union, the implementation process is overseen by the European Commission, ensuring that MS adheres to EU directives. When a directive is adopted, EU MS is obligated to incorporate its provisions into their national legal systems. The Commission monitors this process, including the transposition, application, and enforcement of directives, through regular reports submitted by EU MS (Krämer, 2014).

Under Article 258 of the Treaty on the Functioning of the European Union (TFEU), in cases where an EU MS is found to have violated EU environmental laws, the Commission can initiate formal procedures and, with the EU Court of Justice's endorsement, declare the infringement. The EU MS is then required to take corrective measures. The Commission also has the authority to request financial penalties if an EU MS fails to comply (Krämer, 2014). This system, which includes the ability to impose financial penalties, serves as an effective means of enforcing EU environmental laws, ensuring that regulations related to the CE model are effectively implemented and upheld.

The EU's enforcement mechanism provides a strong framework for upholding environmental regulations, particularly those focused on the CE. It demonstrates the EU's commitment to ensuring that its directives are not only adopted but also correctly implemented at the national level. The ability to impose financial penalties adds an additional layer of deterrence to encourage compliance. Overall, this system plays a crucial role in achieving the EU's environmental and CE objectives, ultimately contributing to a more sustainable and resource-efficient future.

As previously highlighted, when considering the distinctive characteristics that distinguish ASEAN from the EU—such as the principles of state sovereignty and noninterference embodied in the "ASEAN Way," the prevalence of political and nonbinding instruments, and the substantial differences in political systems and socioeconomic development among the AMS—consistently developing a CE model becomes an intricate challenge. Moreover, given that the majority of AMS, apart from Singapore, are categorized as developing nations (WorldData), with Cambodia and Lao PDR even classified as least developed countries (UNCTAD), where a linear economic model remains prevalent, the task becomes even more formidable. Additionally, integrating the development of a CE model as a mandatory requirement in the association's binding policies introduces an added layer of complexity to this challenge.

ASEAN has the opportunity to gain significant insights from the European Union's successful journey toward achieving a CE. This potential learning experience centers on three crucial areas: addressing disparities among AMS, establishing legally binding frameworks, and ensuring effective implementation. Given the urgent environmental challenges, particularly those associated with plastic waste, decisive action from ASEAN is now a necessity. The aspiration is that adopting the EU as a role model will provide valuable guidance to ASEAN in formulating and executing an impactful CE strategy. In recent times, ASEAN has actively collaborated with the EU, seeking to learn best practices in developing a CE. This collaboration is evident in initiatives such as the ASEAN Circular Economy Stakeholder Platform, modeled after the EU Circular Economy Stakeholder Platform, and the EU-funded Enhanced Regional EU-ASEAN Dialogue Instrument (E-READI), facilitating the exchange of experiences and policy dialog on CE between the EU and ASEAN (ASEAN, 2022; EU, 2022).

In light of these instrumental cooperative programs, despite ASEAN not reaching the same level of integration as the EU, it has significantly derived inspiration from the EU's approaches to structuring regional cooperation and integration in addressing gaps between AMS. While achieving a level of integration akin to the EU, ASEAN can draw valuable lessons from the EU experience by enhancing and reinforcing national efforts through the formulation and implementation of initiatives such as regional guidelines on circularity in plastics use, research and innovation on plastics, and technical standards for plastic products and recycled plastics (Akenji et al., 2019, p. 34).

The establishment of regional CE guidelines focusing on plastics' production, utilization, and post-use management holds the potential to disseminate valuable policy insights and best practices among AMS, drawing from experiences in different regions, encompassing the EU. The target audience encompasses governments at different levels, businesses, and stakeholders. Initial topics include strategies such as plastic avoidance and alternative selection, optimization of plastics choices, design principles for a CE of plastics, applications for compostable plastics, safety assessments for food packaging with recycled plastics, addressing hazards related to plastic additives, waste disposal, collection and sorting systems, and appropriate reuse or recycling systems tailored to the ASEAN region (Akenji et al., 2019, p. 35).

Proposing the establishment of an extensive ASEAN research and innovation network focused on sustainable polymers, packaging, and circularity to bolster regional expertise and encourage innovation. The network, with diverse membership from academia, business, governments, and stakeholders, would address alternatives to petroleum-based plastics, business models reducing single-use plastics, and innovative plastic recycling methods. Creating this network necessitates identifying AMS willing to lead and, therefore, promoting good practices as role models for other AMS to adopt in their national contexts. ASEAN can establish a regional knowledge hub on CE and plastics, acting as a coordinating entity for the network to collate data on materials such as plastics and related impacts, promote good practices, and review experiences from similar initiatives globally. For instance, the newly instituted "Circular Materials Lab" in Singapore could serve as an inspiring model (Akenji et al., 2019, pp. 35–36).

5 Conclusion

In contrast to the linear economy, the CE offers multifaceted advantages spanning environmental, economic, and societal dimensions. Nevertheless, it is imperative to underscore that the CE is not a universal panacea and should not supplant the overarching objectives of sustainable development.

AMS, guided by their vision for vibrant, sustainable, and highly integrated economies, recognize the need to depart from wasteful linear consumption patterns.

Instead, the CE model, which emphasizes waste reduction, recycling, and resource efficiency, offers a promising avenue to address the plastic waste crisis while promoting economic growth and environmental preservation.

This chapter has undertaken a comparative analysis of CE policies and regulations in the EU and the ASEAN region, aiming to draw valuable lessons and implications for advancing CE practices in ASEAN. The critical lesson drawn from the EU's CE model is the importance of having robust institutions with binding legal frameworks and effective monitoring systems. The EU's CE policies have evolved beyond mere theoretical concepts; they are now ingrained in the daily lives of its citizens, with practices like waste separation becoming commonplace. The EU has managed to successfully integrate environmental concerns into the fabric of its society.

However, it is essential to recognize that achieving the same level of CE implementation in the ASEAN region is a formidable task. The EU's extensive experience and well-established framework have evolved over decades, making it a high bar to reach. Nonetheless, taking inspiration and learning from the EU's journey toward a more circular and environmentally conscious economy are a valuable approach for ASEAN. By adapting best practices and implementing effective policies, ASEAN can take meaningful steps toward adopting CE principles, even if it may not reach the same level of maturity as the EU in the near term.

References

ACESP. (2024). The ASEAN Circular Economy Stakeholder Platform (ACESP). In *ASEAN circular economy stakeholder platform*. https://ce.acsdsd.org/. Accessed 10.03.24.

Aggeri, F. (2020). The circular economy: Historical perspective and contemporary issues. In K. Delchet-Cochet (Ed.), *Circular economy: From waste reduction value creation* (pp. 3–12). Wiley.

Akenji, L., Bengtsson, M., Kato, M., et al. (2019). *Circular economy and plastics: A gap-analysis in ASEAN member states. European Commission Directorate general for environment and directorate general for international cooperation and development*. Association of Southeast Asian Nations (ASEAN).

Antico, C. (2020). *Brief overview of the EU waste law and policy*.

ASEAN. (1967). *Charter of the Association of Southeast Asian Nations*.

ASEAN. (2019a). *Bangkok Declaration on Combating Marine Debris in ASEAN Region. Bangkok, Thailand*.

ASEAN. (2019b). *ASEAN Framework of Action on Marine Debris. Bangkok, Thailand*.

ASEAN. (2021a). ASEAN adopts framework for Circular Economy. In *Association of Southeast Asian Nations - A community of opportunities for all*. https://asean.org/asean-adopts-framework-for-circular-economy/. Accessed 10.03.24.

ASEAN. (2021b). *Framework for Circular Economy for the ASEAN Economic Community*.

ASEAN. (2022). From linear to circular: EU, ASEAN inaugurate Stakeholder Platform Secretariat to boost region's circular economy. In: *Association of Southeast Asian Nations - A Community of Opportunities for All*. https://asean.org/from-linear-to-circular-eu-asean-inaugurate-stakeholder-platform-secretariat-to-boost-regions-circular-economy/. Accessed 10.03.24.

ASEAN. (2025). *ASEAN Community Vision 2025*.

ASEAN Secretariat. (2021). *ASEAN regional action plan for combating marine debris in the ASEAN member states (2021–2025)*. ASEAN Secretariat.

Barrowclough, D., & Birkbeck, C. D. (2022). Transforming the global plastics economy: The role of economic policies in the global governance of plastic pollution. *Social Sciences, 11*, 1–26.

Bengtsson, M., & Schroder, P. (2021). *Building Back better in ASEAN countries: Opportunities to advance a circular economy*. Hanns Seidel Foundation - Vietnam Office.

Bengtsson, M., Schroder, P., & Siegner, M. (2021). *Building back better through circular economy - opportunities for ASEAN countries*.

Boer, B. (2016). Environmental law in Southeast Asia. In P. Hirsch (Ed.), *Routledge handbook of the environment in Southeast Asia* (1st ed., pp. 115–132). Taylor & Francis Group.

CMS Law. (2023). *Plastic and packaging laws in Germany*.

De Römph, J. T. (2018). *The legal transition towards a circular economy "EU environmental law examined"*. KU Leuven and Hasselt University.

De Wit, W., Burns, E. T., Guinchard, J.-C., & Ahmed, N. (2021). *Plastics: The costs to society, the environment and the economy*. World Wildlife Fund (WWF).

Directorate-General for Environment. (2022a). *Proposal for a revision of EU legislation on Packaging and Packaging Waste*.

Directorate-General for Environment. (2022b). *Call for Evidence: Revision of the Waste Framework Directive*.

Directorate-General for Environment. (2023). *Brochure on EU action against microplastic pollution*.

Dunphy, S. (2020). *Plastic for recycling from Europe ends up in Asian waters*. European Scientist.

Ellen MacArthur Foundation. (2022). *The EU's Circular Economy Action Plan*.

EU. (2022). Adopting circular economy practices requires more than just reducing the use of plastics. In: *EU-ASEAN Strategic Partnership 2022*. https://euinasean.eu/adopting-circular-economy-practices-requires-more-than-just-reducing-the-use-ofplastics/#:~:text=ECONOMY%20PRACTICES%20REQUIRES-,MORE%20THAN%20JUST%20REDUCING%20THE%20USE%20OF%20PLASTICS,prevents%20the%20generation%20of%20waste. Accessed 10.03.24.

EU. (2023). *The ASEAN Circular Economy Platform is now a reality*. https://circulareconomy.europa.eu/platform/en/news-and-events/all-news/asean-circular-economy-platform-now-reality. Accessed 10.03.24.

EU Monitor. (2021). *Circular economy: Commission takes action against five Member States to improve waste management*.

EU-ASEAN Business Council. (2020). *Advancing ASEAN's circular economy: Agenda advocacy paper*. EU-ASEAN Business Council.

European Commission. (2018). *Communication from the Commission to the European Parliament, the Council, the European Economic and Social Committee of the regions "Closing the loop - An EU action plan for the Circular Economy"*.

European Commission. (2020a). *Strategic Plan 2020–2024 DG ENVIRONMENT*.

European Commission. (2020b). *Commission Delegated Regulation (EU) 2020/2174 amending Annexes IC, III, IIIA, IV, V, VII and VIII to Regulation (EC) No 1013/2006 of the European Parliament and of the Council on shipments of waste*.

European Commission. (2021a). *EU restrictions on certain single-use plastics*.

European Commission. (2021b). *Plastic waste shipments*.

European Commission. (2022). *European Green Deal: Putting an end to wasteful packaging, boosting reuse and recycling*.

European Commission. (2023). *Protecting environment and health: Commission adopts measures to restrict intentionally added microplastics*.

European Commission Circular economy action plan.

European Commission Plastic bags.

European Environment Agency. (2022). *Early warning assessment related to the 2025 targets for municipal waste and packaging waste - Country profile: Germany*.

European Environmental Bureau. (2019). *Priorities of the European Green Deal and the 8th Environment Action Programme.*
European Environmental Bureau. (2020). *#NoTimeToWaste member states delay meeting the inevitable targets.*
European Parliament, European Council. (2008). *Directive (EU) 2008/98/EC of the European Parliament and of the Council on waste and repealing certain Directives.*
European Parliament, European Council. (2015a). *Directive (EU) 2015/720 of the European Parliament and of the Council amending Directive 94/62/EC as regards reducing the consumption of lightweight plastic carrier bags.*
European Parliament, European Council. (2015b). *Consolidated text: European Parliament and Council Directive 94/62/EC of 20 December 1994 on packaging and packaging waste.*
European Parliament, European Council. (2018). *Directive (EU) 2018/851 of the European Parliament amending Directive 2008/98/EC on waste.*
European Parliament, European Council. (2019). *Directive (EU) 2019/904 of the European Partiament and of the Council on the reduction of the impact of certain plastic products on the environment.*
Eurostat. (2021). *EU recycled 41% of plastic packaging waste in 2019.*
Farmer, A. (2020). Developing the circular economy in the European Union. In S. K. Ghosh (Ed.), *Circular economy: Global perspective* (pp. 389–412). Springer.
Federal Ministry for the Environment, Nature Conservation, Nuclear Safety, and Consumer Protection (BMUV). (2023). *Waste Management in Germany 2023 - Facts, data, figures.*
Fischer, C. (2013). *Municipal waste management in Germany.*
Ghosh, S. K. (2020). *Circular economy: Global perspective.* Springer.
Government of Vietnam. (2022). *Decree No. 08/2022/ND-CP dated 10 January 2022 detailing a Number of Articles of the Law on Environmental Protection.*
Igini, M. (2022). *How waste management in Germany is changing the game.*
Institute for European Environmental Policy. (2010). *Sourcebook on EU Environmental Law.*
Jain, A. (2020). *Trash trade wars: Southeast Asia's problem with the World's waste.* Council on Foreign Relations.
Jambeck, J. R., Geyer, R., Wilcox, C., et al. (2015). Plastic waste inputs from land into the ocean. *Science, 347*, 768–771. https://doi.org/10.1126/science.1260352
Kálmán, A. (2023). *In numbers: Europe's mounting plastic waste problem unpacked.* Investigate Europe.
Koh, K. L., & Bhullar, L. (2011). Governance on adaptation to climate change in the Asean region. *CCLR, 5*, 82–90.
Koh, K. L., & Robison, N. A. (2002). Strengthening sustainable development in regional intergovernmental governance: Lessons from the "ASEAN way.". *Singapore Journal of International & Comparative Law, 6*, 640–682.
Kopnina, H., & Poldner, K. (2021). Introduction: Challenges and opportunities of the circular economy. In H. Kopnina & K. Poldner (Eds.), *Challenges and opportunities for ethical and sustainable business* (pp. 1–16). Taylor & Francis Group.
Krämer, L. (2014). *EU enforcement of environmental laws: From great principles to daily practice - Improving citizen involvement.*
Library of Congress. (2022). *Germany: Lightweight Plastic Bag Ban to Take Effect January 1, 2022.*
Maggio, A. R. (2019). Regional cooperation and marine environmental protection in Southeast Asia. *Asia-Pacific Journal of Ocean Law and Policy, 4*, 202–220.
Mahaseth, H., & Subramaniam, K. (2021). Binding or non-binding: Analysing the nature of the Asean agreements. *ICLR, 21.*
Mazur-Wierzbicka, E. (2021). Circular economy: Advancement of European Union countries. *Environmental Sciences Europe, 33*, 111. https://doi.org/10.1186/s12302-021-00549-0
Meijer, L. J. J., Emmerik, T., van der Ent, R., et al. (2021). More than 1000 rivers account for 80% of global riverine plastic emissions into the ocean. *Science Advances, 7*, 1–13.

National Assembly of Vietnam. (2020). *Law No. 72/2020/QH14 dated 17 November 2020 on Environmental Protection.*

Oberpriller, S. (2021). EU bans 10 most common single-use plastic items found on beaches. *Euronews.green.*

OECD. (2022). *Global plastics outlook: Economic drivers, environmental impacts and policy options.* OECD.

Okafor, J. (2023). *Which countries are best in low-zero waste management?*

RKC-MPD Vietnam: Action Plans and Road Maps. In: *Regional Knowledge for Marine Plastic Debris (RKC-MPD).* https://rkcmpd-eria.org/framework-regulations/framework-viet-nam/#:~:text=In%202009%2C%20Viet%20Nam%20adopted,use%20of%20landfills%20by%202050. Accessed 10.03.24.

Tangwanichagapong, S., Logan, M., & Visvanathan, C. (2020). Circular economy for sustainable resource management: The case of packaging waste sector in Thailand. In S. K. Ghosh (Ed.), *Circular economy: Global perspectives* (pp. 353–387). Springer.

The European Files. (2022). *Circular economy in Europe: Promoting sustainable products and consumption.*

UNCTAD UN list of least developed countries. In: *UNCTAD Prosperity for all.* https://unctad.org/topic/least-developed-countries/list. Accessed 10.03.24.

UNEP. (2023). *Turning off the tap. How the world can end plastic pollution and create a circular economy.* United Nations Environment Programme (UNEP).

UNEP Our planet is choking on plastic. https://www.unep.org/interactives/beat-plastic-pollution/?gclid=CjwKCAjw7c2pBhAZEiwA88pOF_AVpFlSN1GDcivNoqgc7hscEpQzKfdWkaCKInl-eG9RfStWPpsxMhoC8-UQAvD_BwE. Accessed 10.03.24.

United Nations. (2020). *European Commission's "Circular Economy Action Plan".*

Vu, H. D., Pham, T. G., & Nguyen, T. X. S. (2021). Vietnam's regulations to prevent pollution from plastic waste: A review based on the circular economy approach. *Journal of Environmental Law, 33*, 137–166.

Whitehead, R. (2018). *Asia bearing the brunt of European waste as recycling problems hit crisis point.* FoodNavigator Asia.

Witoonchart, P., & Thampanishvong, K. (2023). Circular economy for ASEAN's sustainable future. *ASEANFocus*, 18–19.

World Bank. (2021a). *Market study for Thailand: Plastics circularity opportunities and barriers.* World Bank Group.

World Bank. (2021b). *Market study for Malaysia: Plastics circularity opportunities and barriers.* World Bank Group.

World Bank. (2021c). *Market study for Philippines: Plastics circularity opportunities and barriers.* World Bank Group.

World Bank. (2021d). *Market study for Vietnam: Plastics circularity opportunities and barriers.* World Bank Group.

World Economic Forum. (2020). *Plastics, the circular economy and global trade.* World Economic Forum.

World Population Review Plastic Pollution by Country. 2023. In *World Population Review.* https://worldpopulationreview.com/country-rankings/plastic-pollution-by-country. Accessed 10.03.24.

WorldData Developing Countries. In: *WorldData.info.* https://www.worlddata.info/developing-countries.php. Accessed 10.03.24.

A Systematic Review of E-Waste Estimation Methods and Models Used in SAARC and ASEAN Countries

Prajyot P. Gaonkar

Abstract E-waste or electronic waste is a waste generated from the disposal of electrical and electronic equipment (EEE). A report titled "Global E-waste Monitor 2020" estimates total e-waste generation in the world as 53.6 million metric tons (MMt), which is expected to cross 74 MMt by 2030. Electrical and electronic equipment are of different types and kinds, some of which are commonly and much more frequently used, and some are not. For this study, the South Asian Association for Regional Cooperation (SAARC) and Association of Southeast Asian Nations (ASEAN) member countries are considered. Preferred Reporting Items for Systematic Reviews and Meta-Analyses (PRISMA) guidelines are followed for this paper. The methodological section includes three key stages which are identification, screening, and filtering the literature. The study is done at three broad levels of analysis, i.e., publication specific, method/model specific, and comparative analysis. Publication specific analysis shows some of the key trends and traits in the publications. Method/model specific analysis highlights the aspects related to methods used, their requirements, and other factors. The comparative analysis shows the disparity and commonality among SAARC and ASEAN. Suggestions and future research directions are discussed in the last section of the paper.

Keywords E-waste · Waste electrical and electronic equipment · WEEE · E-waste estimation · PRISMA · ASEAN · SAARC

1 Introduction

Electronic waste (e-waste) is waste generated from disposal of electrical and electronic equipment (EEE). Electronic waste, e-waste, and waste electrical and electronic equipment (WEEE) are the words which are used interchangeably by different researchers and literatures. India's E-waste (Management) Rule 2022 defines

P. P. Gaonkar (✉)
Goa University, Taleigao Plateau, Goa, India
e-mail: economics.prajyot@unigoa.ac.in

A. Heshmati, P. K. Nam (eds.), *Circular Economy and its Implementations in Southeast Asia*, Frontiers in South and Southeast Asian Development Research,
https://doi.org/10.1007/978-981-96-0539-2_3

e-waste as "Electrical and Electronic Equipment, including solar photo-voltaic modules or panels or cells, whole or in part discarded as waste, as well as rejects from manufacturing, refurbishment and repair processes."

According to the European Union, "waste electrical and electronic equipment" or "WEEE" means electrical or electronic equipment which is waste within the meaning of Article 3(1) of Directive 2008/98/EC, including all components, subassemblies, and consumables which are part of the product at the time of discarding. It categorizes electrical and electronic equipment into ten broad categories, viz., large household appliances, small household appliances, IT and telecommunication equipment, consumer equipment and photovoltaic panels, lighting equipment, electrical and electronic tools, toys, leisure sports equipment, medical devices, and monitoring and control instruments and automatic dispensers (Directive *2012/19/ EU of the European Parliament and of the Council of 4 July 2012 on Waste Electrical and Electronic Equipment (WEEE) (Recast) Text with EEA Relevance*, 2012).

A report titled "Global E-Waste Monitor 2020" estimates total e-waste generation in the world as 53.6 million metric tons (MMt) in the year 2019, which is expected to cross 74 MMt by 2030. An individual generates an average 7.3 kg of e-waste annually in the world. Out of the total estimated e-waste, only 17.4% (9.3 MMt) is formally recycled. At continental level, Asia is the major emitter of e-waste in the world accounting for 24.9 MMt which is 46.45% of world's e-waste generation. Asia's formal e-waste recycling is only 11.7% (2.9 MMt) of its generation. According to the same report, Southeast Asian countries are generating 3.5 MMt of e-waste, and South Asian countries account for 4.8 MMt (Forti et al., 2020).

Another striking feature of an updated report by the same organization titled "Global Transboundary E-Waste Flow Monitor 2022" highlights that there is 5.1 MMt of transboundary e-waste flow detected of which 1.8 MMt is controlled transboundary movement and 3.3 MMt is through uncontrolled channels in the world. The transboundary trade of e-waste within Asia is reported to be quite high. Among Asian region, Southeast Asia and East Asia are the largest importers with 1.1 MMt and 1 MMt of e-waste, respectively. Southern Asia is also a big contributor to e-waste generation, but the import and exports are barely 0.3 MMt and 0.2 MMt, respectively, signifying the presence of internal dismantling facilities especially informal ones (Baldé et al., 2022).

Electronic Waste is not merely waste – it contains significant potential in the form of valuable materials found in waste electronic and electrical equipment. According to (Namias, 2013), there are 60 different elements found in electronic equipment which are ranging from common metals (like copper, aluminum, iron, etc.) to precious metals (like gold, silver, palladium, etc.) and to hazardous metals (like lead, mercury, etc.). The presence of valuable material like gold makes the obsolete EEEs even more valuable. A study by Zeng et al. (2022) found that the recovered value often exceeds the cost of recovery. Apart from positive value, these EEEs are also accompanied with harmful materials. These heavily cost humans as well as environment. Grant et al. (2013) identified various effects on human of exposure to e-waste, like problem related to growth, reproduction, respiration, and mental health.

The paper is classified into six sections. Section 1 deals with introduction, and Sect. 2 covers previous research work and review of literature on the e-waste quantification. The methodological part is discussed in Sect. 3. Section 4 presents result and finding. Section 5 deals with discussion on derived results, and lastly Sect. 6 concludes.

2 Literature Review

2.1 Review of Existing Literature

Yedla (2016) cites various studies estimating e-waste using different methods, some of which includes material flow analysis (MFA), market supply method (MSM), consumption use method, econometric analysis, Carnegie Mellon model, time step method, etc. Among the abovementioned methods, some tries to estimate e-waste from supply side (e.g. market supply method), whereas some tries to evaluate from demand side (e.g., consumption use method); on the other hand, some focuses on macro variables like GDP, growth rate, population, and use of regression tools (e.g., econometric analysis), and some tries to evaluate at micro level with more detailed understanding of flow throughout its lifetime (e.g., material flow analysis) (Yedla, 2016).

In existing literatures, there are dozens of studies available using different methods to estimate electronic waste. Dwivedy and Mittal (2010a) is one of such, which uses a modified theoretical model of Peralta and Fontanos developed in 2006. Another study by Rahmani et al. (2014) used a slightly modified model of Peralta and Fontanos. They included logistic growth model to estimate future waste of electronic equipment. Nigerian researchers used the Carnegie Mellon model to estimate electronic waste in Nigeria's Lagos metropolis city (Ibrahim et al., 2013). A study by Andarani and Goto (2014) estimated e-waste with the help of Steubing's model with some modifications.

Jain and Sareen estimates e-waste in Indian context based on market supply of electronic equipment which is termed as "market supply method" (Jain & Sareen, 2006). Consumption and use method is another popular method which is used by (Singh et al., 2018) in their study of estimating e-waste in the city of Chandigarh, India. Similarly consumption and use method is used in other studies as well (Alavi et al., 2015; Araújo et al., 2012; Iqbal et al., 2020). Another less popular model is population balance model, used by Santoso et al. (2019) to estimate e-waste generation in Indonesia.

There are review papers published on estimating e-waste at the global level as well as at regional level. Ismail and Hanafiah (2020) is one of such kind which reviews available literature on e-waste estimation in different parts of the world. They reviewed publications from across the world and found that most of the included studies used traditional methods of estimating e-waste. On the basis of their finding, one of the suggestions drawn is to use more mathematically sophisticated methods and models.

There is another study by Andeobu et al. (2021), which reviews research publications dealing with e-waste in the Asia-Pacific region. It includes countries like Australia, China, India, Indonesia, and Malaysia. This study is much broader in nature wherein all related aspects to e-waste are considered and reviewed. It includes legislative aspects, quantification aspect, management of e-waste, environmental and health impacts, and problems and challenges of e-waste recycling. It also draws valuable suggestions including the increasing need of enacting laws on e-waste, using modern and new strategies of e-waste management like 3R and extended producer responsibility (EPR), the need of support from local authorities, and the need for international integration.

2.2 Research Motivation and Objectives

Various studies are identified estimating electronic waste using various methods mentioned above. In a study published by (Ikhlayel, 2016), there are conflicting estimations identified by using different methods considering a single nation and timeframe. These disparities and conflicting estimations are caused by the variables used and assumptions made in the applied methodology.

The thorough review of literature raised some questions. Firstly, why are there so many methods developed and used to estimate e-waste? Secondly, if there arises disparity in calculation of quantity at a given time and space, then can we rely on any single method to estimate electronic waste? Thirdly, most of the studies are developed in advanced countries, which puts a question of what is the status of e-waste estimation in developing and less developed countries? With these question in consideration, we aimed our study to identify existing literature in the quantification of e-waste in developing and underdeveloped countries. And secondly, we objectivize our study to evaluate the nature of used methods.

We chose SAARC and ASEAN, two of the biggest regional cooperative organizations in Asia, as our representative of developing and underdeveloped countries. SAARC stands for "South Asian Association for Regional Cooperation" (SAARC), and it includes eight member countries, viz., Afghanistan, Bangladesh, Bhutan, India, Maldives, Nepal, Pakistan, and Sri Lanka (About SAARC, n.d.). ASEAN stands for "Association of Southeast Asian Nations" (ASEAN), and it includes ten member countries, viz., Brunei, Cambodia, Indonesia, Lao PDR, Malaysia, Myanmar, Philippines, Singapore, Thailand, and Vietnam (Member States, n.d.). Both the regional cooperations are objectivized toward their own distinct regional cooperation and development. The choice for SAARC and ASEAN is motivated by shared features of member countries including land and water boundaries, cultural traits, and economic outlooks.

3 Materials and Methods

3.1 Collection of Existing Studies

PRISMA guidelines are used to start the process of systematic literature review (PRISMA, n.d.-a). It is one of the key methods of reviewing literature especially used in medical sciences. There are many evidences which show the use of PRISMA in other fields (Andeobu et al., 2021; Maphosa & Maphosa, 2020). Collection process starts with identification of sources of information. To source research papers, we used Scopus as a key database. Scopus is a widely used database which includes studies from diverse fields. The second step involves the development of search strategy.

Under developing search strategy, the study area needs to be specified. We restricted our search results from SAARC and ASEAN member countries. The shared features and commonalities between SAARC and ASEAN member countries make it a perfect choice to suit our objectives. Followed by the identification of keywords, for that matter some of the existing studies were referred (including Arain et al., 2022; Araújo et al., 2012; Bogar et al., 2019; Ismail & Hanafiah, 2020; Jain & Sareen, 2006). Keywords related to e-waste, estimation or quantification, method or model, and the member countries of SAARC and ASEAN are included. Identified keywords were searched in the title, abstract, and keywords section of the article. The combined set of identified keywords are fed into Scopus advanced search, and results are downloaded during the first week of August 2023. Results were downloaded in a document of CSV file format for further analysis. "Citation information, bibliographic information, abstract and keywords, and funding details" are downloaded for further analysis.

3.2 Selection of Existing Studies

The selection process started using the PRISMA flow diagram (PRISMA, n.d.-b) which is shown in Fig. 1. The PRISMA flow diagram starts with identification of studies which were later screened. Studies which cleared all screening tests are included for final analysis. Screening is done at three key levels, beginning with document type, followed by screening for relevance and lastly screening for eligibility.

Papers with document type "articles," "conference papers," and "reviews" were included. The relevance of study is tested by screening title and abstract. And eligibility was tested using thorough evaluation of the research paper and its compliance with specified eligibility criteria. Studies estimating electronic waste using estimation models or methods are considered as eligible for inclusion.

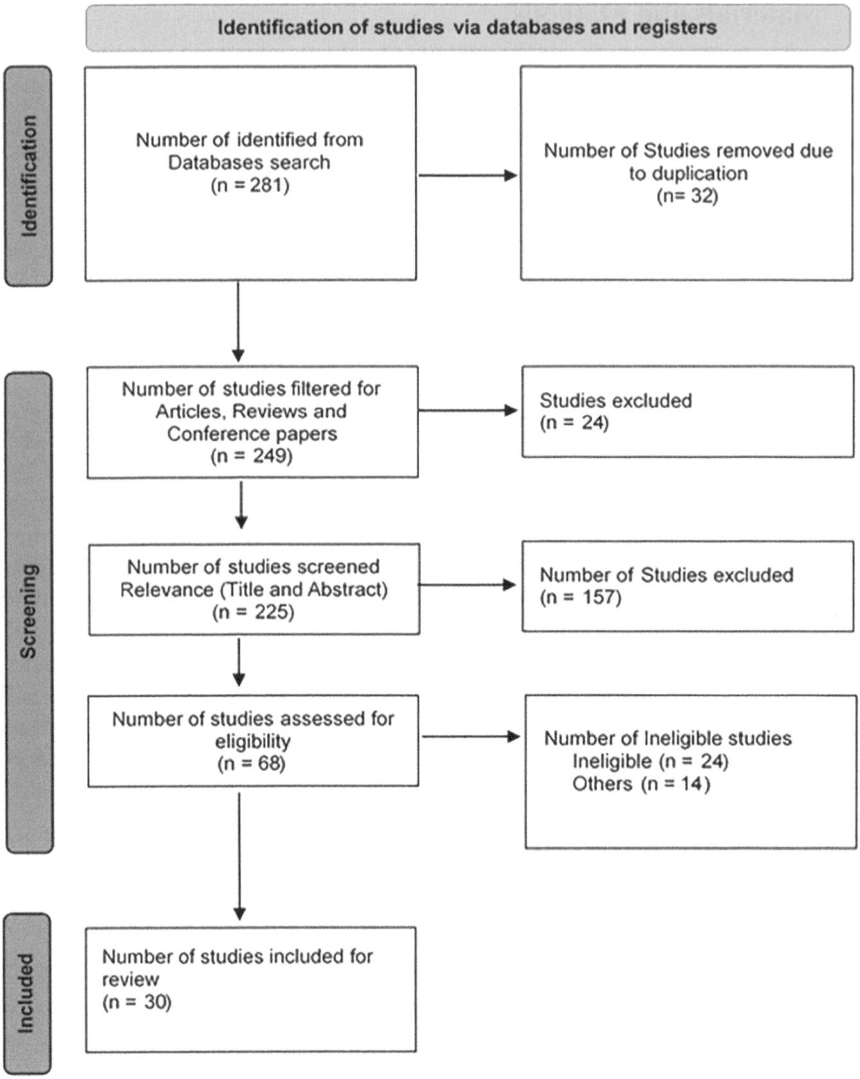

Fig. 1 PRISMA flow diagram showing literature review selection results

3.3 *Analytical Process*

The evaluation and analytical process is classified into three sections, that is, publication specific analysis, method/model specific analysis, and comparative analysis. The publication specific analysis is concerned with analyzing publication related aspects including publishing years, region, document type, EEE types, etc.

The main purpose of this section is to understand variation and trend in the publications. The second section deals with method/model specific analysis, wherein methodological aspects are evaluated including used methods or models, lifespan modeling techniques, prospect of estimation, etc. And at the end, a section on comparative analysis is dedicated to compare and analyze aspects related to publications and methods in both the regional groups, i.e., SAARC and ASEAN.

4 Results and Findings

4.1 Literature Search Results

As shown in Fig. 1, altogether 281 studies were identified, using different search criteria and keywords of which 32 are duplicate and so are removed. Out of the remaining 249, studies with document type as "article," "review," and "conference paper" were included which left us with 225 studies for further evaluation. The remaining studies were screened for relevance on the bases of title and abstract, which left us with only 68 studies for eligibility assessment, eliminating 157 studies. Eligibility assessment is done based on specified inclusion and exclusion criteria. Out of 68 studies, 24 were ineligible and 14 were excluded due to other reasons (i.e., unavailability/inaccessibility). In net, we were left with 30 studies from different regions for the final inclusion. These studies were thoroughly evaluated and reviewed to fulfil the objectives of this review paper.

4.2 Publication Specific Analysis

Figure 2 shows number of selected studies published in different countries of SAARC and ASEAN regional groups. A key observation highlights that India accounts for the highest number of publications studying the generation of e-waste. Indonesia follows the lead with the contribution of five publications, and the remaining countries including Bangladesh, Malaysia, the Philippines, Thailand, and Vietnam account for two to three publications each. It is noticeable that out of 18 member countries, only 7 countries have contribution in estimating e-waste

Year-wise analysis of published literature clearly highlights the increasing trend in publications as shown in Fig. 3. The oldest publication is identified from the year 2005 and the latest is from the year 2022. The number of publications in the last 4 years (2019–2022) is more than that of the total publications from the previous 14 years (2005–2018), clearly highlighting the growing concern over e-waste and its quantification. Out of the total included studies, 23 are articles and 7 are conference papers (Fig. 4).

Studies assessed e-waste at different levels, that is, at the regional, state, or province level and at national level. Studies estimating e-waste of a country or a

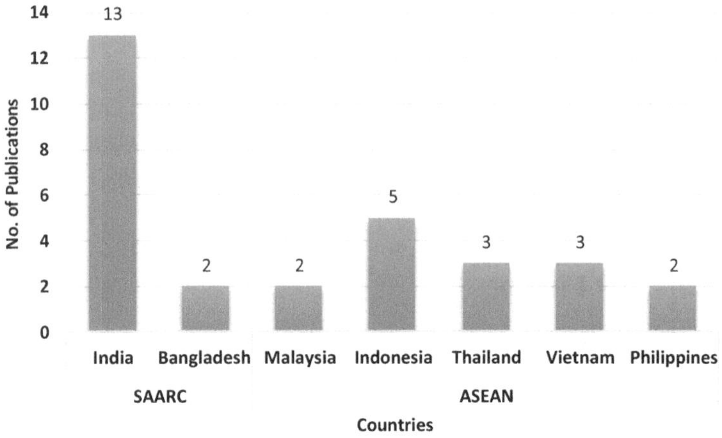

Fig. 2 Number of studies published in SAARC and ASEAN member countries

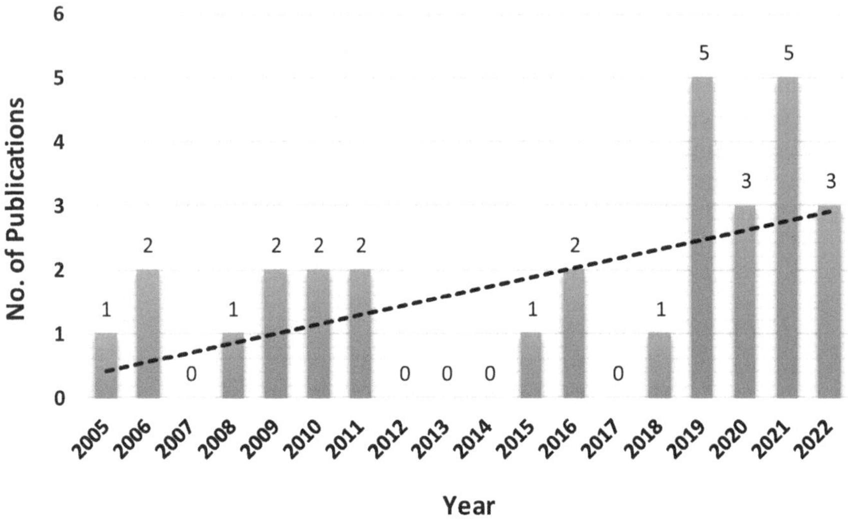

Fig. 3 Year-wise analysis of the published studies

nation are classified under "national" level assessment, studies estimating e-waste for states or provinces are classified under "state/province" level assessment, and studies estimating at micro or individual level including institute, locality, or city are classified under "regional/local" level. Figure 5 shows the count of studies conducted at different levels. Assessment at national level is highest with 69% (20 studies) of the studies—followed by regional level with 28% (9 studies) of studies and state level with only 3% (1 study) of studies.

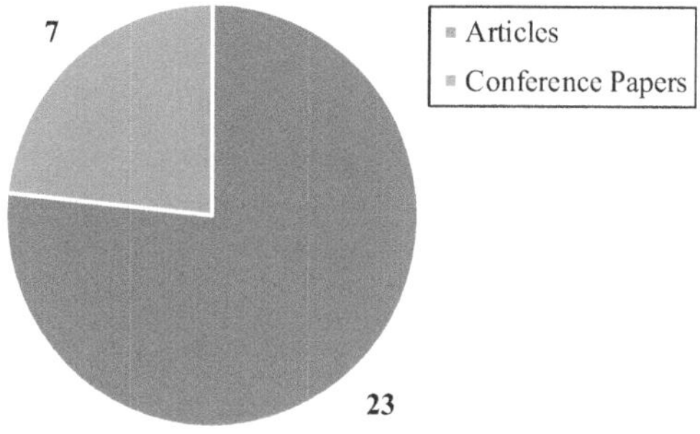

Fig. 4 Classification of studies on the bases of document type

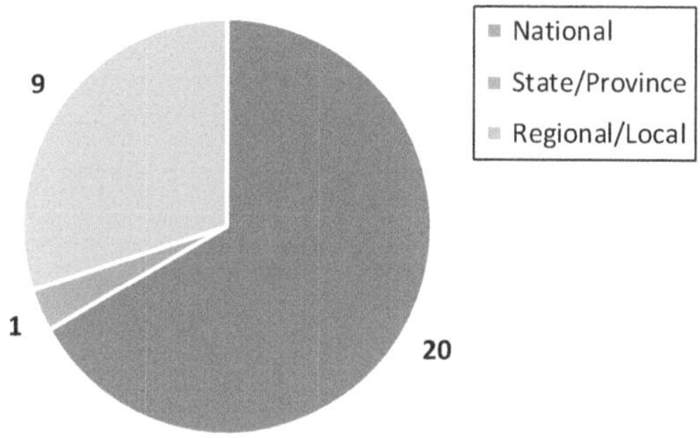

Fig. 5 Assessment at different levels

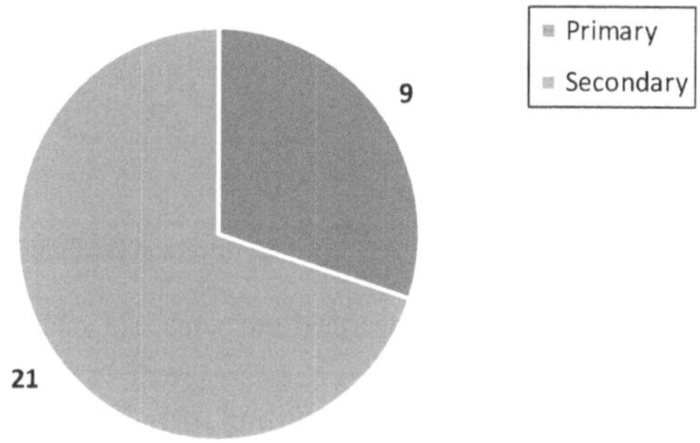

Fig. 6 Studies objectivizing e-waste estimation as primary and secondary goal

We could also find diversity in the quantification of e-waste as an objective of the study. Most studies were conducted with the sole objective of estimation of e-waste or development of a new method/model of e-waste estimation, and the remaining were conducted considering estimation as an accompanying objective for the fulfilment of other primary objectives. Among included studies, 70% (21 studies) considers e-waste estimation as primary objective, whereas the remaining 30% (9 studies) considers estimation of e-waste as an accompanying objective, as shown in Fig. 6.

Estimation could also be done for single or multiple items. Figure 7 shows the proportion of studies choosing EEEs for estimation, wherein around 37% of the studies quantifies e-waste generation for single EEE, whereas the remaining 63% of

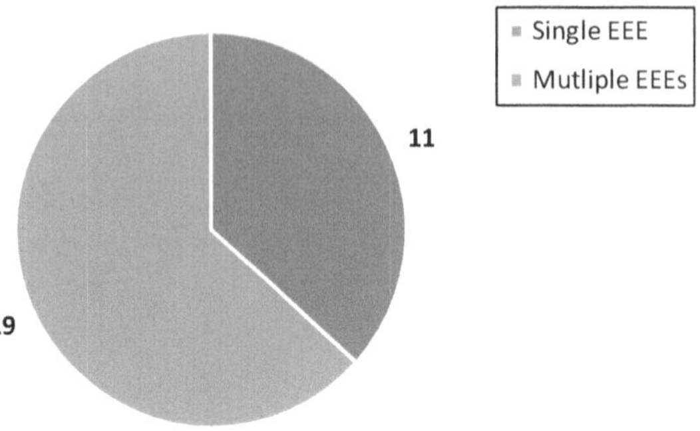

Fig. 7 Single and multiple EEE considered for estimation

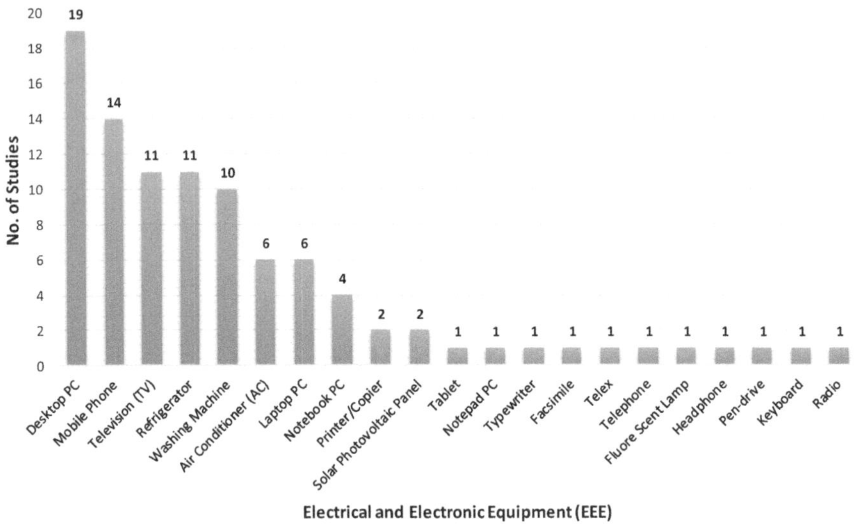

Fig. 8 Different EEEs identified and number of studies evaluating each EEE

the studies quantifies for multiple EEEs. Even in some studies, close to ten EEEs are assessed at a time (Iqbal et al., 2022; Ismail & Hanafiah, 2019; Panchal et al., 2021). Among the selected EEEs for estimation, desktop PC, mobile phones, television, refrigerator, and washing machine are the majorly chosen equipment. There are 21 different EEEs evaluated in the selected publications—these primarily include IT and telecommunications devices, small household appliances, and large household appliances, as identified in Figs. 8 and 9.

4.3 *Method/Model Specific Analysis*

Table 1 contains data matrix of method/model specific aspect of the studies. In all the selected studies, there are 17 unique methods/models of e-waste estimation

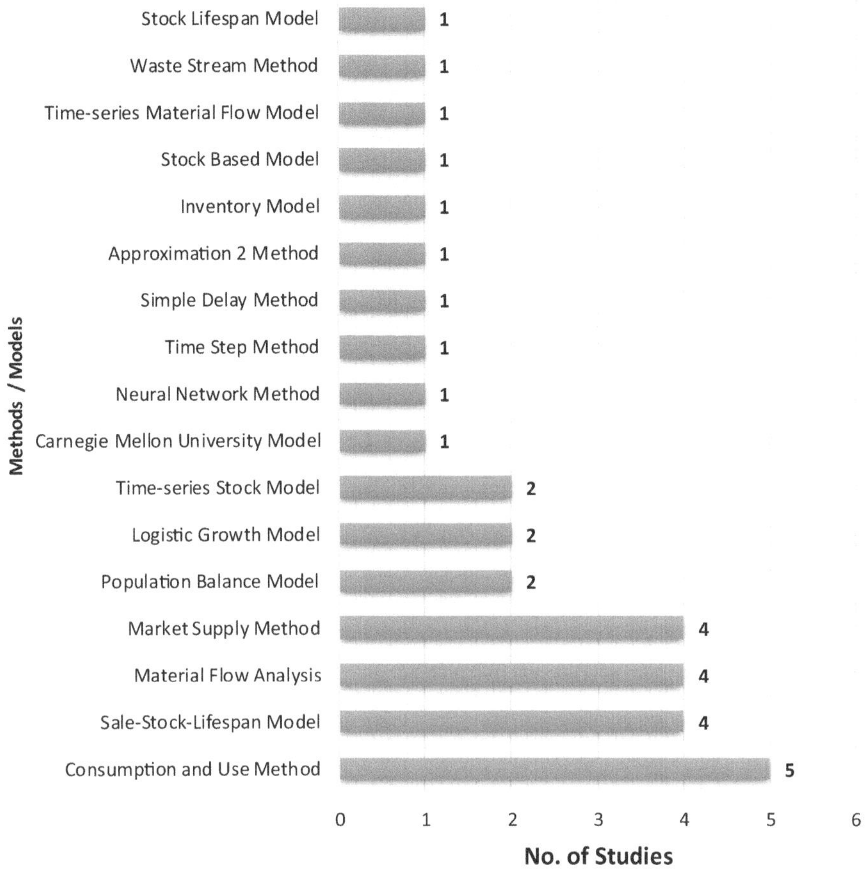

Fig. 9 Identified e-waste estimation method/model and their frequency of use

identified. Among these, "consumption and use method" (Iqbal et al., 2019; Ismail & Hanafiah, 2019; Nguyen et al., 2009; Siringo et al., 2019), "market supply method" (Jain, 2008; Jain & Sareen, 2006; Panchal et al., 2021; Streicher-Porte et al., 2005), "material flow analysis" (Agamuthu et al., 2015; Andarani & Goto, 2014; Mairizal et al., 2021; Yedla, 2016), and "sales-stock-lifespan model" (Faircloth et al., 2019; Gautam et al., 2021; Koshta et al., 2021; Sakthivel et al., 2022) are most commonly used.

These methods/models require various types and kinds of data (see Table 1). Among these data requirements, sales, lifespan/lifetime of the equipment, and weight are the commonly demanded ones. These data requirements are mostly sourced from secondary sources, but some studies collected these parameters using primary data collection tools as well. Lifespan/lifetime of the equipment can be considered without any modeling techniques which will resemble a single assumed "use-time" of the equipment. Modeled lifespan/lifetime gives much more statistical

Table 1 Model specific analytical data matrix

Papers	Method/model used	Data requirements for estimation	Lifespan modeling technique	Method for future estimate extrapolation (if future estimate)
Kiran et al. (2021)	Stock lifespan model	Sales data, lifespan data, average weight	Weibull distribution function	Multiple regression model
Ismail and Hanafiah (2019)	Consumption and use method	Sales data, lifespan data, stock data (penetration data)		
Iqbal et al. (2019)	Consumption and use method	Sales data, lifespan data, average weight, stock data (penetration data)		Linear trend forecasting technique
Jain (2008)	Market supply method	Sales data, lifespan, average weight		
Siringo et al. (2020)	Stock based model	Sales data, lifespan	Weibull distribution function	
Siringo et al. (2019)	Consumption and use method	Sales data, stock data, lifespan		
Yedla (2016)	Material flow analysis	Sales data, discard flow fraction, lifespan		
Panchal et al. (2021)	Market supply method	Sales data, lifespan, average weight		
Kunacheva et al. (2009)	Inventory model	Sales data, lifespan, average weight, replacement factor		
Iqbal et al. (2022)	Consumption and use method, waste stream method	Sales data, lifespan data, average weight, stock data (penetration data)		Linear trend forecasting technique
Mairizal et al. (2021)	Material flow analysis	Sales data, discard flow fraction, lifespan, average weight		Linear trend forecasting technique
Gautam et al. (2021)	Sales-stock-lifespan model	Installed capacity, failure rate (lifespan)	Weibull distribution function	Secondary data
Santoso et al. (2019)	Population balance model	Sales data, stock data, lifespan	Weibull distribution function	Linear trend forecasting technique
Budnard and Khaodhiar (2022)	Approximation 2 method, simple delay method, time step method	Sales data, lifespan, average weight		
Koshta et al. (2021)	Sales-stock-lifespan model	Sales data, stock data, lifespan		

(continued)

Table 1 (continued)

Papers	Method/model used	Data requirements for estimation	Lifespan modeling technique	Method for future estimate extrapolation (if future estimate)
			Weibull distribution function	
Dwivedy and Mittal (2010a)	Sale data, lifespan, discard flow path			
Kosai et al. (2020)	Population balance model	Sales data, lifespan	Weibull distribution function	Brass model
Tran et al. (2016)	Time-series stock model	Sales data, lifespan	Normal distribution	Linear trend forecasting technique
Galang and Ballesteros Jr. (2018)	Neural network method			
Jain and Sareen (2006)	Market supply method	Sales data, lifespan, average weight		
Agamuthu et al. (2015)	Material flow analysis	Sales data, discard flow fraction, lifespan, average weight		
Peralta and Fontanos (2006)	Carnegie Mellon University model (developed version of material flow analysis)	Sales data, discard flow fraction, lifespan, average weight		
Vivek et al. (2019)	Time-series stock model	Sales data, lifespan		Seasonal auto regressive integrated moving average (ARIMA) model
Jain and Chawla (2014)	Logistic growth model	Sales data, penetration rate, lifespan		Logistic growth
Dwivedy and Mittal (2010b)	Logistic growth model	Sales data, penetration rate, lifespan		Logistic time-series growth
Streicher-Porte et al. (2005)	Market supply method	Sales data, lifespan		
Andarani and Goto (2014)	Material flow analysis	Sales data, discard flow fraction, lifespan, average weight		
Sakthivel et al. (2022)	Sales-stock-lifespan model	Sales data, stock data, lifespan		

(continued)

Table 1 (continued)

Papers	Method/model used	Data requirements for estimation	Lifespan modeling technique	Method for future estimate extrapolation (if future estimate)
			Weibull distribution function	
Faircloth et al. (2019)	Sales-stock-lifespan model	Sales data, stock data, lifespan	Weibull distribution function	
Nguyen et al. (2009)	Consumption and use method	Sales data, lifespan data, average weight, stock data (penetration data)	Weibull distribution function	

Notes: semicolons (,) are used for separations

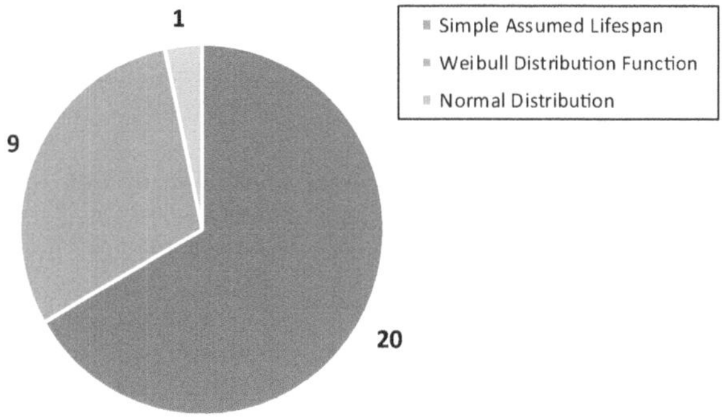

Fig. 10 Number of studies using different lifespan modeling techniques

accuracy for estimation. For that matter, various lifespan modeling techniques are available, among which "Weibull distribution function" is the most frequently used method. Nine out of ten studies applying statistical lifespan modeling tools are witnessed using Weibull distribution function. A great chunk of studies is identified using single assumed lifespan for EEE which are either sources from primary or secondary sources (Fig. 10).

Measurement units also have a significant effect on the estimated volume. There are two key units of measurement identified, that is, e-waste is either measured in number of physical units or weight. Around half of the identified studies uses only "number of units" as the unit of measurement, i.e., 52% of the studies—only "weight" as a unit of measurement is used by 27% of the studies, and the remaining 21% of studies quantifies in both the units, i.e., "number of units" as well as "weight" (see Fig. 11).

Figure 12 shows the proportion of studies estimating current and future flow of e-waste. Current estimation is the estimation of obsolete EEEs which are already in use. However, future estimation is for that quantity which is yet to be put in market. Among the studies, only 38% account for estimating future e-waste generation,

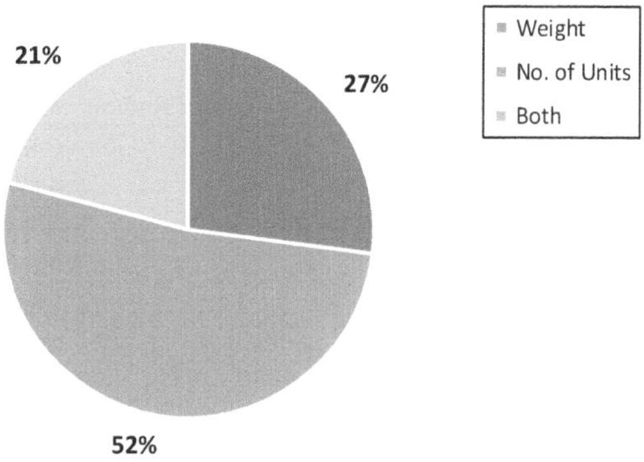

Fig. 11 Percentage of studies using different units of measurement

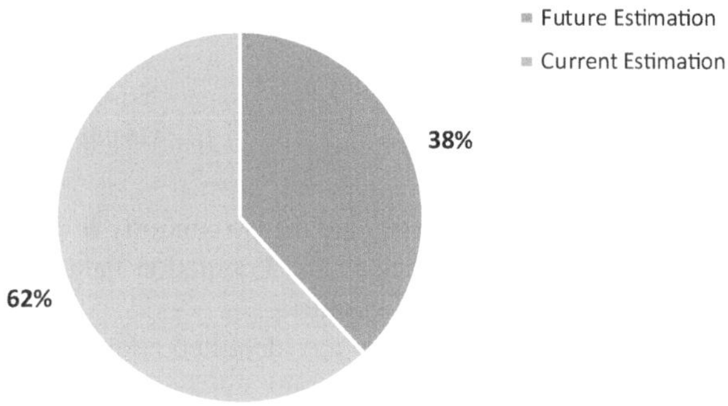

Fig. 12 Percentage of studies estimating future and current flow of e-waste

while 62% estimates current flow of e-waste. To get future estimation of the EEE waste stream, different techniques and methods are used, as shown in Table 1. Among these, logistic growth model and linear trend forecasting techniques are the most used ones.

4.4 Comparative Analysis

The selected regional groups have membership of 18 countries out of which 8 belong to SAARC and 10 belong to ASEAN. India and Bangladesh are the only nations whose contribution in e-waste estimation is included from the SAARC region. The combined count of studies from both these countries accounts for half of the total selected studies. From the ASEAN region, Malaysia, Indonesia, the Philippines, Thailand, and Vietnam are the nations from which studies estimating e-waste are identified and included (see Fig. 2).

Table 2 shows the use of different methods/models in SAARC and ASEAN regions. From Table 2, it is evident that "market supply method" and "sales-stock-lifespan model" are majorly used methods for estimation of e-waste in SAARC countries with four and three studies, respectively. "Consumption and use method"

Table 2 Number of different methods/models used

SAARC		ASEAN	
Method/model	No. of studies	Method/model	No. of studies
Stock lifespan model	1	Consumption and use method	3
Consumption and use method	2	Stock based model	1
Market supply method	4	Material flow analysis	3
Material flow analysis	1	Inventory model	1
Waste stream method	1	Sales-stock-lifespan model	1
Sales-stock-lifespan model	3	Population balance model	2
Time-series material flow model	1	Approximation 2 method	1
Time-series stock model	1	Simple delay method	1
Logistic growth model	2	Time step method	1
		Time-series stock model	1
		Neural network method	1
		Carnegie Mellon model	1

and "material flow analysis" are the two majorly used estimation methods/models in ASEAN with three studies each. "Population balance model" is another emerging model evident in ASEAN countries.

In SAARC, seven studies are identified addressing future estimation of e-waste flow wherein there is wide diversity evident in the use of techniques of forecasting the future flow. On the other hand, five studies are identified in the ASEAN region using "linear trend forecasting technique" as the future forecasting method (see Table 1).

In ASEAN region, assessment is done either at "national level" or "regional/local level," wherein 87% of studies identified estimating e-waste at national level and 13% at regional/local level. In the SAARC region, along with other two, there is an inclusion of state/province level assessment. Out of the total, 50%, 43%, and 7% of the studies are identified evaluating e-waste at national, regional/local, and state/province level, respectively, in SAARC region (Fig. 13).

Figure 14 clearly highlights the distinction in the use of measurement units in both the regions. "Number of units" as a unit of measurement is widely used in both the regions accounting for eight and seven studies in ASEAN and SAARC, respectively. Only "weight" as a unit of measurement is found to be used most frequently in SAARC as five studies are identified, whereas in ASEAN, only three studies are identified using "weight" as sole measurement unit. From both the regions, studies are also identified estimating and reporting e-waste in both the units—among these, the ASEAN has four studies and SAARC has two studies.

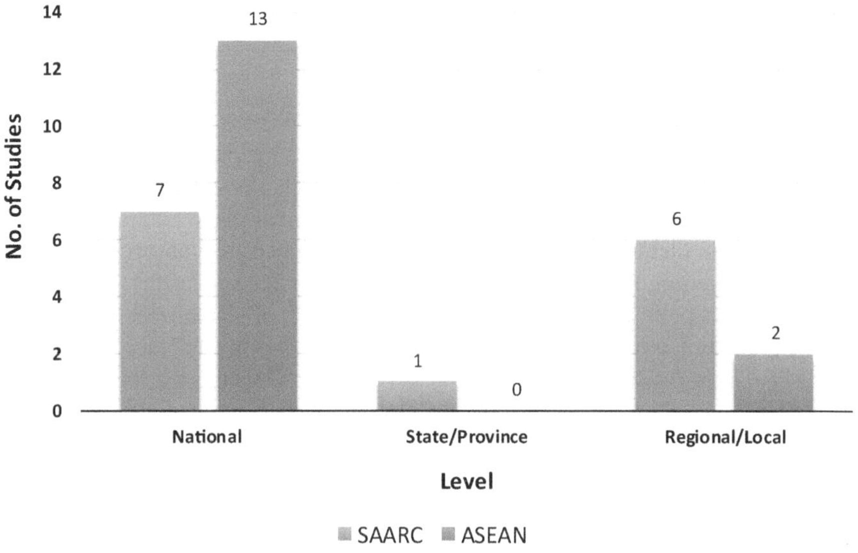

Fig. 13 Number of studies conducted at different levels in SAARC and ASEAN countries

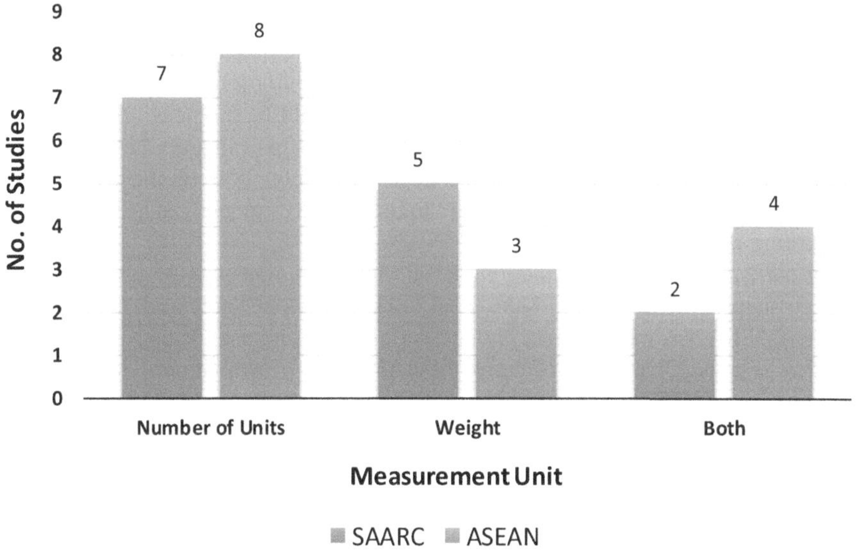

Fig. 14 Regional distinction in unit of measurement

5 Discussion and Suggestions

5.1 *Current Literature Progress*

The selected studies (30 studies) were from 7 out of 18 member nations from both the regions. India accounts for the larger part of contribution where 13 studies are identified. Year-wise evaluation shows an increasing trend in the publications. The spiking phase in literature and interest among researchers can be observed in recent years, that is, during the past 4 years (2019–2022). Major part of the studies was research articles published in different journals and most of which were considering e-waste estimation as the primary objective. It is clearly visible that most studies are assessing e-waste estimation at national level. International level evaluation is ignored by the selected studies, but studies from other parts of the world has shown interest to evaluate at international level as highlighted by Ismail and Hanafiah (2020). More than half of the studies preferred quantifying e-waste generation for multiple EEE—among the most common are daily use equipment like mobile phones, TVs, desktop PCs, and washing machines.

Nonstatistical method which relies on user and market information is widely used for estimating e-waste. Most of the methods rely heavily on secondary data, especially sales data. The lifespan is in fact modeled to get more statistical accuracy in estimation, for which many statistical techniques are used like Weibull distribution function. Around half of the studies accounted e-waste quantity in physical units. Quantification in physical units does not serve greater purpose when further analysis is concerned for which measurement in "weight" could be the better choice. Estimation of future flow of e-waste gives a proper direction to evaluators and authorities about the threats and required policy steps. But only 38% of the studies are estimating future flow of e-waste using "logistic growth model" and "linear trend forecasting technique."

Disparity in the use of method/model in the two regions can be identified—"-market supply method' and "sales-stock-lifespan model" are used in SAARC, whereas ASEAN is much keener about "consumption and use method" and "material flow analysis." In terms of current and future flow estimation, SAARC have literature contribution with equal proportion, but variation in ASEAN could be observed (see Fig. 15). Studies from SAARC use a variety of future forecasting techniques, but ASEAN studies are centered at only two major types. In SAARC, most of the studies conducted their assessment at national level, whereas in ASEAN, national level consideration is comparatively lower. In terms of accounting units, we do not observe much difference in measuring physical units, but there are comparative differences in terms of weight and both as the units of measure.

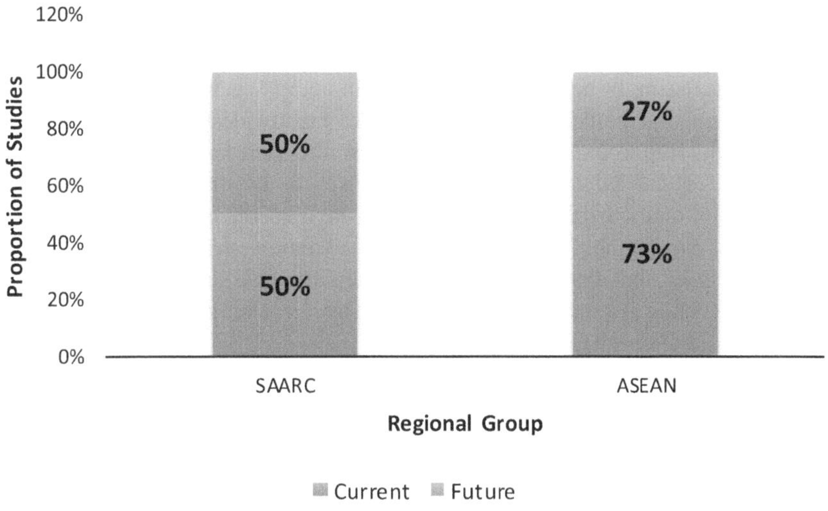

Fig. 15 Current and future estimation in SAARC and ASEAN

5.2 *Future Research Directions*

Future research and literature in e-waste estimation need to be much broader and inclusive in nature. There are lot of other electrical and electronic equipment put in market which can be harmful alongside being valuable if recovered—these needs attention. Secondly, using more statistically accurate models gives better accuracy in quantification. Thirdly, evaluation of future flow of e-waste is needed for better policy outcomes. Fourthly, using weight as a measuring unit for studies with multiple EEEs will serve a better purpose in accountability. And lastly, a combined efforts could be put by researchers and authorities from these member countries to develop a much sophisticated model for estimating e-waste according to the nature and requirements of the region.

This study is restricted only to SAARC and ASEAN member countries which poses a limitation for the current study to account and consider estimation methods in other parts of Asian and other developing countries. Future research could be done at a global level considering other developing countries as well. The second limitation is posed using databases. Scopus has pretty much good content of literature, but including literature from other databases may serve a greater purpose. Further research could be diversified at a much deeper level including detailed analysis of data sources and with proper detailed diagnostics of methods/models.

6 Conclusion

E-waste is increasingly being a threat to human society and the environment. An effective estimation could largely help to take rightly designed policy actions. E-waste estimation around the world and across different countries with different characteristics is carried out by authorities as well as by many researchers using varied forms of methods/models. These models are used according to the needs and conditions of the region. There is a plenty of literature present on estimation in different parts of the world. A concerning literature has been identified which questions the validity of methods/models used. This demands the need for evaluation of e-waste estimation methods.

This paper tried to identify e-waste estimation methods in SAARC and ASEAN member countries, as both are seen closely linked in terms of share features. PRISMA guidelines and flow diagram are followed for the systematic review of literature. Presented findings show data on methods/models used in e-waste estimation and identify disparities and commonalities among SAARC and ASEAN. The identified literature shows some stunning facts about the developing nations and their attitude toward e-waste estimation. Interest in e-waste accounting has developed recently. Out of 18, only 7 member countries are identified with their contribution in e-waste estimation. There are wide variety of methods identified across different member countries. Assessment clearly highlights the differences in used methods. Diversity in accounting units is also visible. Estimating the future flow of e-waste is neglected. Further research in estimation of e-waste can be undertaken at much broader and much scientific levels.

Acknowledgments The author acknowledges the guidance received from Prof. Sudhakar Yedla, Professor from India Gandhi Institute of Developmental Research (IGIDR) Mumbai, India, and thanks the institute for their opportunity to work in the campus. The author's acknowledgment also extends to the efforts of Prof. Pranab Mukhopadhyay, Professor in Goa University, Goa, India. The principal of Government College of Arts, Science, and Commerce, Khandola, Marcela, Goa, India, is also well acknowledged for the support.

References

About SAARC. (n.d.). Retrieved August 5, 2024, from https://www.saarc-sec.org/index.php/about-saarc/about-saarc

Agamuthu, P., Kasapo, P., & Mohd Nordin, N. A. (2015). E-waste flow among selected institutions of higher learning using material flow analysis model. *Resources, Conservation and Recycling, 105*, 177–185. https://doi.org/10.1016/j.resconrec.2015.09.018

Alavi, N., Shirmardi, M., Babaei, A., Takdastan, A., & Bagheri, N. (2015). Waste electrical and electronic equipment (WEEE) estimation: A case study of Ahvaz City, Iran. *Journal of the Air & Waste Management Association, 65*(3), 298–305. https://doi.org/10.1080/10962247.2014.976297

Andarani, P., & Goto, N. (2014). Potential e-waste generated from households in Indonesia using material flow analysis. *Journal of Material Cycles and Waste Management, 16*(2), 306–320. https://doi.org/10.1007/s10163-013-0191-0

Andeobu, L., Wibowo, S., & Grandhi, S. (2021). A systematic review of E-waste generation and environmental management of Asia pacific countries. *International Journal of Environmental Research and Public Health, 18*(17), 9051. https://doi.org/10.3390/ijerph18179051

Arain, A. L., Neitzel, R. L., Nambunmee, K., Hischier, R., Jindaphong, S., Austin-Breneman, J., & Jolliet, O. (2022). Material flow, economic and environmental life cycle performances of informal electronic waste recycling in a Thai community. *Resources, Conservation and Recycling, 180*, 106129. https://doi.org/10.1016/j.resconrec.2021.106129

Araújo, M. G., Magrini, A., Mahler, C. F., & Bilitewski, B. (2012). A model for estimation of potential generation of waste electrical and electronic equipment in Brazil. *Waste Management, 32*(2), 335–342. https://doi.org/10.1016/j.wasman.2011.09.020

Baldé, C. P., D'Angelo, E., Luda, V., Deubzer, O., & Kuehr, R. (2022). *Global Transboundary E-waste flows monitor—2022*. United Nations Institute for Training and Research (UNITAR).

Bogar, Z. O., Capraz, O., & Güngör, A. (2019). An overview of methods used for estimating E-waste amount. In *Electronic waste management and treatment technology* (pp. 53–75). Elsevier. https://doi.org/10.1016/B978-0-12-816190-6.00003-0

Budnard, J., & Khaodhiar, S. (2022). Estimating the generation of discarded Mobile phones and highlighting areas for recycling precious metals from printed circuit boards in Thailand. *Sustainability, 14*(24), 17025. https://doi.org/10.3390/su142417025

Directive 2012/19/EU of the European Parliament and of the Council of 4 July 2012 on waste electrical and electronic equipment (WEEE) (recast) Text with EEA relevance. (2012). http://data.europa.eu/eli/dir/2012/19/oj/eng

Dwivedy, M., & Mittal, R. K. (2010a). Estimation of future outflows of e-waste in India. *Waste Management, 30*(3), 483–491. https://doi.org/10.1016/j.wasman.2009.09.024

Dwivedy, M., & Mittal, R. K. (2010b). Future trends in computer waste generation in India. *Waste Management, 30*(11), 2265–2277. https://doi.org/10.1016/j.wasman.2010.06.025

Faircloth, C. C., Wagner, K. H., Woodward, K. E., Rakkwamsuk, P., & Gheewala, S. H. (2019). The environmental and economic impacts of photovoltaic waste management in Thailand. *Resources, Conservation and Recycling, 143*, 260–272. https://doi.org/10.1016/j.resconrec.2019.01.008

Forti, V., Baldé, C. P., Kuehr, R., & Bel, G. (2020). *The Global E-waste Monitor 2020.*

Galang, M. G. K., & Ballesteros, F., Jr. (2018). Estimation of waste Mobile phones in The Philippines using neural networks. *Global NEST Journal, 20*(4), 767–772. https://doi.org/10.30955/gnj.002534

Gautam, A., Shankar, R., & Vrat, P. (2021). End-of-life solar photovoltaic e-waste assessment in India: A step towards a circular economy. *Sustainable Production and Consumption, 26*, 65–77. https://doi.org/10.1016/j.spc.2020.09.011

Grant, K., Goldizen, F. C., Sly, P. D., Brune, M.-N., Neira, M., Van Den Berg, M., & Norman, R. E. (2013). Health consequences of exposure to e-waste: A systematic review. *The Lancet Global Health, 1*(6), e350–e361. https://doi.org/10.1016/S2214-109X(13)70101-3

Ibrahim, F. B., Adie, D. B., Giwa, A.-R., Abdullahi, S. A., & Okuofu, C. A. (2013). *Material flow analysis of electronic wastes (e-Wastes) in Lagos, Nigeria. 2013*. https://doi.org/10.4236/jep.2013.49117

Ikhlayel, M. (2016). Differences of methods to estimate generation of waste electrical and electronic equipment for developing countries: Jordan as a case study. *Resources, Conservation and Recycling, 108*, 134–139. https://doi.org/10.1016/j.resconrec.2016.01.015

Iqbal, M., Al Nadim, A., Ahmed, H., & Iqbal, S. (2019, August 1). *Estimation of E-waste generation in Sylhet City and its management.*

Iqbal, M., Siddique, F., & Biswas, D. (2020). *E-waste estimation: A case study of Shahjalal University of Science and Technology, Sylhet-3114, Bangladesh.*

Iqbal, M., Siddique, F., Biswas, D., & Shamsuzzoha, A. (2022). Electronic waste generation by selected students of a University in Bangladesh: A case study. *Journal of Hazardous, Toxic, and Radioactive Waste, 26*(4), 05022001. https://doi.org/10.1061/(ASCE)HZ.2153-5515.0000719

Ismail, H., & Hanafiah, M. M. (2019). *A preliminary study on e-waste generation from households in Malaysia*. 060011. https://doi.org/10.1063/1.5111273.

Ismail, H., & Hanafiah, M. M. (2020). A review of sustainable e-waste generation and management: Present and future perspectives. *Journal of Environmental Management, 264*, 110495. https://doi.org/10.1016/j.jenvman.2020.110495

Jain, A. (2008). Application of WEEE/E-waste inventorization methodology using tracer tracking along material flow in EEE lifecycle in developing country. In *2008 IEEE International Symposium on Electronics and the Environment* (pp. 1–4). IEEE. https://doi.org/10.1109/ISEE.2008.4562906

Jain, N., & Chawla, P. (2014). Future outflows of toxic material from end-of-life computers in India. *International Journal of Environmental Technology and Management, 17*(2/3/4), 237. https://doi.org/10.1504/IJETM.2014.061796

Jain, A., & Sareen, R. (2006). E-waste assessment methodology and validation in India. *Journal of Material Cycles and Waste Management, 8*(1), 40–45. https://doi.org/10.1007/s10163-005-0145-2

Kiran, M., Shanmugam, P. V., Mishra, A., Mehendale, A., & Nadheera Sherin, H. R. (2021). A multivariate discrete grey model for estimating the waste from mobile phones, televisions, and personal computers in India. *Journal of Cleaner Production, 293*, 126185. https://doi.org/10.1016/j.jclepro.2021.126185

Kosai, S., Kishita, Y., & Yamasue, E. (2020). Estimation of the metal flow of WEEE in Vietnam considering lifespan transition. *Resources, Conservation and Recycling, 154*, 104621. https://doi.org/10.1016/j.resconrec.2019.104621

Koshta, N., Patra, S., & Singh, S. P. (2021). Estimation of E-waste at micro level for reverse logistics: A case of Delhi. *Journal of Cleaner Production, 314*, 128063. https://doi.org/10.1016/j.jclepro.2021.128063

Kunacheva, C., Juanga, J. P., & Visvanathan, C. (2009). Electrical and electronic waste inventory and management strategies in Bangkok, Thailand. *International Journal of Environment and Waste Management, 3*(1/2), 107. https://doi.org/10.1504/IJEWM.2009.024703

Mairizal, A. Q., Sembada, A. Y., Tse, K. M., & Rhamdhani, M. A. (2021). Electronic waste generation, economic values, distribution map, and possible recycling system in Indonesia. *Journal of Cleaner Production, 293*, 126096. https://doi.org/10.1016/j.jclepro.2021.126096

Maphosa, V., & Maphosa, M. (2020). E-waste management in sub-Saharan Africa: A systematic literature review. *Cogent Business & Management, 7*(1), 1814503. https://doi.org/10.1080/23311975.2020.1814503

Member States. (n.d.). *ASEAN Main Portal*. Retrieved August 5, 2024, from https://asean.org/about-asean/member-states/

Namias, J. (2013). *Obstacles and domestic solutions*.

Nguyen, D.-Q., Yamasue, E., Okumura, H., & Ishihara, K. N. (2009). Use and disposal of large home electronic appliances in Vietnam. *Journal of Material Cycles and Waste Management, 11*(4), 358–366. https://doi.org/10.1007/s10163-009-0264-2

Panchal, R., Singh, A., & Diwan, H. (2021). Economic potential of recycling e-waste in India and its impact on import of materials. *Resources Policy, 74*, 102264. https://doi.org/10.1016/j.resourpol.2021.102264

Peralta, G. L., & Fontanos, P. M. (2006). E-waste issues and measures in The Philippines. *Journal of Material Cycles and Waste Management, 8*(1), 34–39. https://doi.org/10.1007/s10163-005-0142-5

PRISMA. (n.d.-a). Retrieved August 10, 2023, from http://prisma-statement.org/PRISMAStatement/Checklist?AspxAutoDetectCookieSupport=1

PRISMA. (n.d.-b). Retrieved August 10, 2023, from http://prisma-statement.org/prismastatement/flowdiagram.aspx

Rahmani, M., Nabizadeh, R., Yaghmaeian, K., Mahvi, A. H., & Yunesian, M. (2014). Estimation of waste from computers and mobile phones in Iran. *Resources, Conservation and Recycling, 87*, 21–29. https://doi.org/10.1016/j.resconrec.2014.03.009

Sakthivel, U., Swaminathan, G., & Anis, J. J. J. (2022). Strategies for quantifying metal recovery from waste electrical and electronic equipment (WEEE/E-waste) using mathematical approach. *Process Integration and Optimization for Sustainability, 6*(3), 781–790. https://doi.org/10.1007/s41660-022-00250-6

Santoso, S., Zagloel, Y. M., Ardi, R., & Suzianti, A. (2019). Estimating the amount of electronic waste generated in Indonesia: Population balance model. *IOP Conference Series: Earth and Environmental Science, 219*, 012006. https://doi.org/10.1088/1755-1315/219/1/012006

Singh, M., Thind, P. S., & John, S. (2018). An analysis on e-waste generation in Chandigarh: Quantification, disposal pattern and future predictions. *Journal of Material Cycles and Waste Management, 20*(3), 1625–1637. https://doi.org/10.1007/s10163-018-0726-5

Siringo, R., Herdiyansyah, H., Kusumastuti, R. D., & Lucianto, A. E. (2019). Assessing materials from hoarded mobile phones: Hidden e-waste subject for reverse logistics. *IOP Conference Series: Materials Science and Engineering, 703*(1), 012021. https://doi.org/10.1088/1757-899X/703/1/012021

Siringo, R., Herdiyansyah, H., & Kusumastuti, R. D. (2020). Applying a stock-based model for estimating the amount of personal computer waste. *IOP Conference Series: Earth and Environmental Science, 456*(1), 012026. https://doi.org/10.1088/1755-1315/456/1/012026

Streicher-Porte, M., Widmer, R., Jain, A., Bader, H.-P., Scheidegger, R., & Kytzia, S. (2005). Key drivers of the e-waste recycling system: Assessing and modelling e-waste processing in the informal sector in Delhi. *Environmental Impact Assessment Review, 25*(5), 472–491. https://doi.org/10.1016/j.eiar.2005.04.004

Tran, H. P., Wang, F., Dewulf, J., Huynh, T.-H., & Schaubroeck, T. (2016). Estimation of the unregistered inflow of electrical and electronic equipment to a domestic market: A case study on televisions in Vietnam. *Environmental Science & Technology, 50*(5), 2424–2433. https://doi.org/10.1021/acs.est.5b01388

Vivek, L. S., Khuteta, A., & Shaha, S. (2019). Forecasting of scrapped (e-waste) mobiles. In *2019 IEEE 5th international conference for convergence in technology (I2CT)* (pp. 1–5). IEEE. https://doi.org/10.1109/I2CT45611.2019.9033730

Yedla, S. (2016). Development of a methodology for electronic waste estimation: A material flow analysis-based SYE-waste model. *Waste Management & Research: The Journal for a Sustainable Circular Economy, 34*(1), 81–86. https://doi.org/10.1177/0734242X15610421

Zeng, X., Xiao, T., Xu, G., Albalghiti, E., Shan, G., & Li, J. (2022). Comparing the costs and benefits of virgin and urban mining. *Journal of Management Science and Engineering, 7*(1), 98–106. https://doi.org/10.1016/j.jmse.2021.05.002

Part II
Regulatory, Monetary and Financial Policy

A Transmission Mechanism of Monetary Policy and Circular Economy in Vietnam

Tu Duong Quach and Tuan Minh Cao

Abstract This research examines the transmission mechanism of monetary policy and the circular economy in Vietnam from 2000 to 2020, utilizing the Bernanke and Mihov approach and a semi-structured vector autoregression (SVAR) model. The study identifies exogenous shocks to monetary policy beyond the control of the State Bank of Vietnam. The results indicate that monetary easing shocks lead to an economic growth recovery after four quarters but result in a significant increase in the discount rate. Surprisingly, the consumer price index initially shows no response, posing difficulties in monetary transmission. Additionally, the analysis of exchange rate shocks reveals substantial pressure on the domestic currency to depreciate, with nominal exchange rate adjustments displaying signs of overcontrol. The monetary policy index closely associates changes in the State Bank's response with innovations in the exchange rate. This research highlights the intricate dynamics of monetary policy shocks in Vietnam and their impact on the circular economy.

Keywords Monetary policy shock · Policy stance · Semi-structured vector autoregression (SVAR) model · Transmission mechanism · Circular economy

1 Introduction

Vietnam, like many other countries worldwide, confronts the intricate challenge of managing monetary policy to effectively meet both its development objectives and maintain macroeconomic stability. Following the global financial crisis of 2008, Vietnam's economy has been grappling with persistent macroeconomic instability, with high inflation posing a threat to stability and the economy's long-term growth targets.

According to the Law on the State Bank of Vietnam (2010), known as Central Bank of Vietnam, the main task of monetary policy is to stabilize the value of

T. D. Quach · T. M. Cao (✉)
School of Economics, Can Tho University, Can Tho City, Vietnam
e-mail: qdtu@ctu.edu.vn; cmtuan@ctu.edu.vn

A. Heshmati, P. K. Nam (eds.), *Circular Economy and its Implementations in Southeast Asia*, Frontiers in South and Southeast Asian Development Research,
https://doi.org/10.1007/978-981-96-0539-2_4

money, ensure the banking system's safety, and promote economic development. However, it is well established that monetary policy changes can have an impact on inflation (Lucas Jr., 1975), as demonstrated in Vietnam during the period from 2008 to 2015.

Vietnam's accession to the World Trade Organization (WTO) in 2007 resulted in an influx of foreign investment, which compelled the State Bank of Vietnam to increase the money supply to stabilize the exchange rate. This significant foreign investment led to a substantial credit growth rate, soaring to more than 51.39% in 2007. However, despite the high credit growth, the real economic growth rate for the period from 2000 to 2015 averaged around 6.5% annually, while the average inflation rate for the same period stood at 7.5% per annum (General Statistics Office of Vietnam, 2016).

Interestingly, a closer examination of credit growth rates in relation to economic growth and inflation rates suggests a concentration of cash flows in the financial sector rather than the manufacturing sector. According to a report by the General Statistics Office of Vietnam (2016), credit growth in the core economic sectors of industry and agriculture was only 5.99% and 11.32%, respectively. In contrast, other sectors such as construction, logistics, telecommunications, consumption, and real estate experienced significant credit growth rates, jumping as high as 38.16%.

Despite efforts to support growth during low inflation, the International Monetary Fund (2014) found persistent supply disparities. This has led to output growth falling below potential and inflation showing signs of rising. Consequently, the State Bank of Vietnam faces challenges in maintaining its target inflation policy and safeguarding the economy from internal and external shocks.

Considering the aforementioned issues, there are indications that the implementation of monetary policy in Vietnam lacks adequate consideration of transmission effects and has not yielded the desired outcomes. To gain deeper insights into these complexities and explore potential improvements, this study utilizes the semi-structured vector autoregression (SVAR) model to analyze the transmission mechanism of exogenous monetary policy shocks in Vietnam during the period 2000–2020. By doing so, this study aims to contribute to a better understanding of Vietnam's monetary policy framework and its implications for economic development and stability.

Moreover, the transmission mechanism of monetary policy in Vietnam plays a pivotal role in shaping the adoption of circular economy practices. Monetary policy decisions, such as interest rate changes, have a direct impact on investment choices, potentially influencing investments in sustainable technologies or eco-friendly projects. However, it is important to note that certain aspects of monetary policy, like inflation targeting, may inadvertently encourage disposable goods consumption rather than promoting a circular economy.

To strike a balance between economic stability and sustainability, a comprehensive assessment of how monetary policy affects circular economy initiatives in Vietnam is necessary. This study aims to explore this relationship by examining how monetary policy decisions influence investment and consumption behaviors related to circular practices. Ultimately, the goal is to provide insights that can help tailor monetary policy tools to better support the transition toward a more circular and sustainable economy in Vietnam.

2 Literature Review

Economists who primarily rely on data and information from the Federal Reserve (Fed) have pioneered the field of empirical studies on monetary policy. However, Sims (1980) raised concerns about the reliability of studying macroeconomic issues using large-scale statistical models. In response, Sims proposed a vector autoregression (VAR) approach, which demonstrates that monetary policy shocks do not adequately explain cyclical fluctuations in macrovariables. Moreover, the hypothesis of the Phillips curve is also rejected based on the VAR model's results. Subsequent research by Sims (1986) and Christiano and Eichenbaum (1992) confirms these findings as they continue to employ the VAR model for verification, leading to similar outcomes.

Building on the VAR methodology, Christiano and Ljungqvist (1988) incorporate the Granger causality test into their analyses. Bernanke and Blinder (1992) also advocate this approach and argue that Federal Funds Rates serve as a more suitable indicator of monetary policy dynamics. Nevertheless, the aforementioned methods only allow economists to test a limited portion of reality compared to what can be visually observed through the dataset (Friedman, 1995), despite the validity of VAR or Granger causality models. Furthermore, Bernanke (1986) highlights the lack of robustness in using VAR to establish a connection between money and income or to distinguish short-term factors in real and nominal price shocks.

To address these challenges, Bernanke (1986) proposes employing orthogonal structural modeling of the residuals of the VAR model. This method proves to be beneficial in offering a clearer explanation of the relationship between money supply and income. Subsequently, Bernanke and Blinder (1992) and Bernanke and Mihov (1998) followed a similar approach with macrovariables to develop semi-structured models for measuring the stance of monetary policy using data from the Federal Reserve. The changes in measurement methods yield more accurate estimates than the traditional approaches.

Drawing inspiration from Fung and Yuan (2000), who focused on measuring monetary policy in Canada, Amarasekara (2008) adopts the semi-structured vector autoregression (SVAR) approach with similar macrovariables, employing datasets from Sri Lanka to gauge the stance of monetary policy. The findings of this study add to the credibility and dependability of the SVAR method.

The primary objective of this study is to uncover the fundamental structural connections between monetary policy measures and macroeconomic fluctuations in Vietnam. To achieve this goal, the study adopts the SVAR methodology. By employing this approach, the study can conduct a thorough and detailed analysis of the interactions among crucial economic variables. In essence, the SVAR methodology provides a more comprehensive and nuanced assessment of the complexities involved in the transmission of monetary policy within the Vietnamese context. Consequently, it offers valuable insights into how monetary policy influences Vietnam's circular economy.

Keynes (1913) initially introduced promotional objectives in the context of central banking, primarily observed in the statutes of central banks in developing and emerging economies. In contrast, central banks in advanced economies often informally convey such objectives. Despite this, central banks possess powerful tools to influence credit allocation and financial firms' investment behavior. The extent to which a central bank promotes circular economy projects depends on its legal mandate and ability to correct market failures compared to other policy institutions (Dikau et al., 2020).

Monetary policy can significantly impact the transition to a circular economy (Husseina & Hamdanb, 2020). There are three existing funding mechanisms based on unconventional monetary policies targeting green or climate investments, known as "Smart Unconventional Monetary" (SUMO) policies. SUMO is made up of three parts:

(1) Special Drawing Rights (SDRs) from the International Monetary Fund (IMF), which can be used to start an international climate fund for low-carbon projects in developing countries (Ferron & Morel, 2014)
(2) Green Quantitative Easing (Green QE), in which central banks borrow more money to pay for low-carbon projects, which is called a "green recovery" (Murphy & Hines, 2010)
(3) Carbon Certificates, in which central banks sell liquid assets to pay for low-carbon projects (Ferron & Morel, 2014)

These SUMO mechanisms have the potential to provide substantial low-cost funds for green projects, reduce risks for private investors, and generate considerable funding while aligning with quantitative easing policies conducted in the UK and the USA. The scale of funding in most proposals is around $100 billion per year, similar to developed countries' commitments to low-carbon development in developing nations (Husseina & Hamdanb, 2020).

3 Data and Methodology

3.1 Semi-structured Vector Autoregression (SVAR) Model

In this research, the semi-structured vector autoregression (SVAR) method is employed to evaluate the effects and transmission mechanisms of monetary policy shocks in Vietnam. The application of this method draws inspiration from its initial utilization by Bernanke and Mihov (1998), where they elucidated the interplay between policy measures and the broader economy in the following manner:

$$Y_t = \sum_{i=0}^{k} B_i Y_{t-i} + \sum_{i=1}^{k} C_i P_{t-i} + A^y v_t^y \tag{1}$$

$$P_t = \sum_{i=0}^{k} D_i Y_{t-i} + \sum_{i=0}^{k} G_i P_{t-i} + A^p v_t^p \tag{2}$$

where B, C, D, G, A^y, and A^p are the regression coefficient vectors of the two equations, Y represents the non-policy variables, and P is the vector of policy variables. The elements in the v^p vector are the monetary policy shocks, v^s, which may include the money supply and money demand shocks. From Eqs. 1 and 2, we find the link between the macrovariables and the policy shocks v^s by putting the variables on the right-hand side back to one lag, where Eqs. 1 and 2 are rewritten:

$$Y_t = (I - B_0)^{-1} \sum_{i=1}^{k} B_i Y_{t-i} + (I - B_0)^{-1} \sum_{i=1}^{k} C_i P_{t-i} + (I - B_0)^{-1} A^y v_t^y \tag{1'}$$

$$\begin{aligned} P_t = {} & (I - G_0)^{-1} \sum_{i=1}^{k} \left[D_i + D_0 (I - B_0)^{-1} B_i \right] Y_{t-1} + (I - G_0)^{-1} \sum_{i=1}^{k} \\ & \times \left[G_i + D_0 (I - B_0)^{-1} C_i \right] P_{t-1} \\ & + \left[(I - G_0)^{-1} D_0 (I - B_0)^{-1} A^y v_t^y + (I - G_0)^{-1} A^p v_t^p \right] \end{aligned} \tag{2'}$$

Rewrite Eqs. (1′) and (2′) in reduced form:

$$Y_t = \sum_{i=1}^{k} H_i^y Y_{t-i} + \sum_{i=1}^{k} H_i^p P_{t-i} + u_t^y \tag{3}$$

$$P_t = \sum_{i=1}^{k} J_i^y Y_{t-i} + \sum_{i=1}^{k} J_i^p P_{t-i} + \left[(I - G_0)^{-1} D_0 u_t^y + u_t^p \right] \tag{4}$$

Comparing residuals between (4) and (2′), we obtain the following:

$$u_t^p = (I - G_0)^{-1} A^p v_t^p \tag{5}$$

Or,

$$u_t^p - G_0 u_t^p = A^p v_t^p \leftrightarrow u = Gu + Av \tag{6}$$

3.2 A Model to Estimate Monetary Policy Shocks in Vietnam

According to Bernanke and Mihov (1998), Eq. (6) is the standard form of the SVAR model. For Vietnam, the paper used the recursive SVAR method based on Cholesky matrix resolution and the method of defining structural noise (Amarasekara., 2008; Bernanke & Mihov, 1998) to build an experimental model based on Eq. (6). The linkages among the elements in the policy group are described below:

$$u_M + \beta u_{TS} + \alpha u_R = v^m \tag{7}$$

$$u_{TS} = \alpha^m v^m + \alpha^e v^e + \alpha^r v^s + v^{ts} \tag{8}$$

$$u_{ER} + \gamma u_R = \delta^e v^e \tag{9}$$

$$u_R = \phi^m v^m + \phi^{ts} v^{ts} + \phi^e v^e + v^s \tag{10}$$

This study used three policy tools of the State Bank of Vietnam (SBV), including money supply, exchange rate, and interest rates, as these tools are quite similar to the abovementioned research. Accordingly, Eq. (7) describes a change in money demand (reserve money), in which the change in money demand is related to changes in money supply and interest rates and v^m represents the money demand shock. Money supply is determined by Eq. (8), which reacts to money demand shocks (v^m), exchange rate shocks (v^e), and monetary policy shocks (v^s). With a fixed exchange rate regime, when the supply of foreign currencies increases or decreases, the SBV must increase or decrease the corresponding money supply to adjust the exchange rate. Similarly, increasing the money supply will also reduce interest rates and, hence, increase capital inflows into the economy. Equation (9) describes a change in the exchange rate, which is related to changes in interest rates; the money supply is assumed not to affect changes in the exchange rate (Fung & Yuan, 2000); and v^e is the exchange rate shock. Equation (10) describes changes in interest rates, which are made up of shocks, including the monetary policy shock v^s. We rewrite it as a matrix of Eq. (6).

$$\begin{bmatrix} 1 & \beta & 0 & \alpha \\ 0 & 1 & 0 & 0 \\ 0 & 0 & 1 & \gamma \\ 0 & 0 & 0 & 1 \end{bmatrix} \begin{bmatrix} u_M \\ u_{TS} \\ u_{ER} \\ u_R \end{bmatrix} = \begin{bmatrix} 1 & 0 & 0 & 0 \\ \alpha^m & 1 & \alpha^e & \alpha^r \\ 0 & 0 & \delta^e & 0 \\ \phi^m & \phi^{ts} & \phi^e & 1 \end{bmatrix} \begin{bmatrix} v^m \\ v^{ts} \\ v^e \\ v^s \end{bmatrix} \tag{11}$$

Then,

$$v^s = w^M u_M + w^{TS} u_{TS} + w^{ER} u_{ER} + w^R u_R \tag{12}$$

In which,

$$w^M = \frac{\alpha^m \phi^{ts} - \phi^m}{1 - \alpha^r \phi^{ts}}$$

$$w^{TS} = \frac{(\alpha^m \phi^{ts} - \phi^m)\beta}{1 - \alpha^r \phi^{ts}} - \frac{\phi^{ts}}{1 - \alpha^r \phi^{ts}}$$

$$w^{ER} = \frac{\phi^{ts} \alpha^e - \phi^e}{\delta^e (1 - \alpha^r \phi^{ts})}$$

$$w^R = \frac{(\alpha^m \phi^{ts} - \phi^m)\alpha}{1 - \alpha^r \phi^{ts}} + \frac{(\phi^{ts} \alpha^e - \phi^e)\gamma}{\delta^e (1 - \alpha^r \phi^{ts})} + \frac{1}{1 - \alpha^r \phi^{ts}}$$

Building the overidentification model to estimate Eq. (11) requires increasing the number of constraints required by the SVAR model. Therefore, at least one constraint must be added.[1] Consequently, adding more constraints results in the formation of three cases:

Case 1: Under the assumptions of Amarasekara (2008) and Fung and Yuan (2000), interest rates have no interaction with changes in the exchange rate, so $\gamma = 0$. In addition, based on the assumption by Strongin (1995) and empirical research by Bernanke and Mihov (1998), changes in money reserves do not respond to interest rate changes. Accordingly, an increase or decrease in money reserves has implications similar to a change in money supply, and, hence, $\alpha = 0$.

Case 2: Inheriting the assumption in Case 1, Fung and Yuan (2000) found that the lending rate is completely unresponsive to money demand shocks. Instead, the SBV harmonizes shocks to the credit market and the exchange rate. Accordingly, the constraints will include $\phi^m = 0$.

Case 3: The assumption in Case 1 continues to be inherited. To avoid dollarization, the SBV operates the USD interest rate at zero, so the term spread or money supply side will be seemingly kept out of the exchange rate shocks. Therefore, restriction $\alpha^e = 0$ to indicate the operation of the SBV for foreign currency deposit interest in Vietnam.

In addition, according to Bernanke and Mihov (1998), a policy variable vector derived from Eq. 5 can be added to the SVAR model:

$$(A^P)^{-1}(I - G_0)P \tag{13}$$

where P is the vector of policy variables; the formula (13) is also used to calculate the monetary policy index (MPI), with the policy variable being the short-term interest rates.

[1]The minimum number of constraints: $2K^2 - \frac{K(K+1)}{2} = K^2 + \frac{K(K-1)}{2} = 4^2 + \frac{4(4-1)}{2} = 22$.

3.3 *Data Collection*

The study utilizes a quarterly dataset covering the period from 2000:1 to 2020:4, comprising a total of 84 observations. The data is categorized into two groups: the policy group and the non-policy group. Within the policy group, the variables include the money reserve in the economy (M), the commercial lending interest rate (R), the short-term deposit rate (TS), and the nominal VND/USD exchange rate (ER). On the other hand, the non-policy group consists of gross domestic product (GDP), inflation (CPI), and the Ho Chi Minh City Stock Exchange Index (VNI). According to previous studies, VNI's direct impact on the dynamics of monetary policy justifies its inclusion as a macrovariable.

The inclusion of the World Commodity Price Index and the Federal Funds Rate in the econometric model serves a critical purpose in assessing their influence on Vietnam's circular economy. These variables have been incorporated to provide a comprehensive understanding of the multifaceted dynamics at play. Firstly, the World Commodity Price Index helps elucidate the global economic integration aspect of the circular economy. By considering the impact of global commodity price fluctuations, the model examines whether circular businesses in Vietnam may be affected not only by domestic factors but also by international cost dynamics. Secondly, the analysis of the Federal Funds Rate offers insights into policy synchronization between Vietnam and larger economies, demonstrating that economic stability is interconnected. This synchronization is vital for creating a favorable environment for circular initiatives to thrive. External factors like global commodity prices and foreign interest rates can introduce uncertainties that circular businesses must anticipate and address in their strategies to ensure resilience and sustainability within Vietnam's circular economy landscape.

4 Results and Discussions

4.1 *The SVAR Model Results*

Given the constraints of the SVAR model, it is ideal to work with a monthly dataset. However, due to the unavailability of monthly macroeconomic data in Vietnam, a quarterly dataset is utilized. Fortunately, previous research by Bernanke and Mihov (1998) has shown that the conclusions drawn from quarterly data do not significantly differ qualitatively from those derived from monthly data. Additionally, Geweke and Runkle (1995) have demonstrated that time aggregation does not cause problems when identifying monetary policy using econometric methods.

The variables used in the study are transformed into logarithmic form, except for the lending interest rates and the FFR. Stationarity tests and lag calculations show that a lag of six is the best based on the likelihood ratio standard, and the FPE, AIC, and HQIC standards all come to the same conclusion. The unit root test results are reported in Appendix Table 2. Notably, other conventional studies on monetary policy often use monthly datasets and lags ranging from 12 to 18 (equivalent to 4–6 quarters), making the lag of 6 selected in this case a perfect fit for the analysis. In Table 1, the results of the SVAR model estimation are shown for three situations where constraints are used to meet the overidentification model's requirements.

The estimates derived from Table 1 consistently align with each other, despite the differing conditions. The value of β, representing the elasticity of short-term deposit rates concerning money demand, is positive and statistically significant. This indicates that increases in interest rates correspond to a decline in money demand.

The parameters of the money supply function (α^m, α^e, and α^r) are all positive, but only α^e is not significant. The values of α^m indicate that when money demand shocks appear, interest rates will increase to stabilize the credit market. The parameter α^r (mean value) reveals that the term spread will rise 39.2 basis points to respond to the 100 basis points increase in lending rates as the result of two compensation effects of monetary policy shocks: the liquidity effect and the expected-inflation effect, as indicated by Fung and Yuan (2000). Short-term deposit rates helped to counteract the pressure of currency depreciation, but the parameter value is not statistically significant because commercial banks always keep USD deposit rates at 0%.

Parameters δ^e consistently hold the same value in all four cases and demonstrate statistical significance. The value $\delta^e = 0.006$ indicates that a 100-point depreciation pressure on the domestic currency only leads to a 0.60-point change in the exchange rate, signifying a control that results in overvaluation.

Both parameters ϕ^m and ϕ^{ts} exhibit statistical significance in all cases, reflecting the interaction between the SBV and the innovation in money supply and exchange rate shocks, respectively. A positive ϕ^m indicates that during a money demand shock, the economy faces liquidity shortages, leading to an increase in interest rates rather than a decrease to offset liquidity. Similarly, a positive ϕ^{ts} requires increases in VND lending rates in strained financial market conditions, prompting commercial banks to reduce their liquidity risk. The SBV pursues a stable exchange rate target, leading to an increase in interest rates to mitigate currency depreciation shocks, thus yielding $\phi^e > 0$. However, ϕ^e lacks statistical significance, suggesting that policy rules do not overreact to devaluation pressure, money reserve changes, or money supply fluctuations. The reason for this is the obvious fact that legal restrictions have an impact on exchange rate management in Vietnam in addition to monetary instruments. In essence, the approach to exchange rate management in Vietnam involves a combination of both monetary policies and legal regulations. These legal barriers play a significant role in shaping the dynamics of exchange rate interventions, thereby impacting the overall effectiveness and implementation of exchange rate policies within the country. As a result, comprehending the complexities that arise from this dual framework is crucial when delving into the intricacies of Vietnam's exchange rate management system.

Equation 12 involves the weights of the four policy variables—w^M, w^{TS}, w^{ER}, and w^R—which, respectively, represent the weights of money demand, money supply, exchange rate, and interest rate variables. We estimate these four weights using the method described by Amarasekara (2008). We take the average of the estimated

Table 1 Estimated results of SVAR model

	β	α^m	α^e	α^r	δ^e	ϕ^m	ϕ^{ts}	ϕ^e	OIR test
Case 1	0.621 (0.038)	1.599 (0.135)	0.006 (0.124)	0.395 (0.019)	0.007 (0.0006)	1.438 (0.406)	3.339 (0.141)	0.050 (0.403)	509.1 (0.000)
Case 2	0.639 (0.048)	1.140 (0.063)	0.021 (0.124)	0.383 (0.022)	0.007 (0.0006)	0 -	3.445 (0.182)	0.101 (0.414)	510.3 (0.000)
Case 3	0.622 (0.038)	1.600 (0.135)	0 -	0.397 (0.019)	0.007 (0.0006)	1.446 (0.405)	3.326 (0.140)	0.030 (0.034)	509.1 (0.000)
Mean	0.627	1.446	0.014	0.392	0.007	1.442	3.370	0.060	
Weight	w^M	−10.687	w^{TS}	3.796	w^{ER}	5.705	w^R	−3.115	

Note. Values in parentheses are standard errors. As for the OIR test, values in parentheses are *p*-value parameters from all three cases and leave out the constraint values in the model (11). According to Bernanke and Mihov (1998) and Fung and Yuan (2000), this method yields results that are more appropriate than when using a single model.

The analysis of the weights reveals that w^{ER} is positive, indicating that devaluation of the local currency results in a monetary tightening shock. On the other hand, w^{M} is negative, suggesting that an increase in the money supply leads to a shock of easing monetary policy. Moreover, the rise in short-term interest rates (w^{TS}) has a tightening effect on climbing lending rates.

The unexpected alignment of the sign of w^{R} with the observed outcomes may seem puzzling at first, but a closer examination reveals a nuanced explanation. The intricate interplay of various factors comes into play, especially when considering the impact of both the money supply and interest rates on the economy.

The interaction between an increase in the money supply and higher interest rates may initially seem to align with conventional economic wisdom, suggesting that these actions could lead to an easing of monetary policy, ultimately fostering economic expansion. However, in practice, this combination can produce nuanced outcomes. Increasing both the money supply and interest rates at the same time can have the opposite effect of what was intended. This could slow down the growth that was wanted and change how circular economy projects are promoted. This intricacy necessitates thoughtful policy formulation to support circular economy goals effectively.

Within this complex monetary landscape, a broader analysis underscores a consistent trend in recent years. This trend underscores the central role of the SBV in shaping monetary policy, primarily focusing on the regulation of the money supply and interest rates to manage the exchange rate strategically. By adeptly adjusting both the money supply and interest rates in tandem, the central bank aims to exert precise control over the value of the Vietnamese currency relative to other currencies. This strategic approach to exchange rate management serves several pivotal economic objectives.

One key motivation for prioritizing exchange rate management may be to ensure stability in international trade, a vital factor for promoting circular economy activities. A controlled exchange rate offers predictability and assurance to exporters and importers, streamlining cross-border transactions associated with circular goods and materials. Moreover, influencing the exchange rate can enhance the competitiveness of Vietnamese goods and services in global markets, bolstering the country's export-oriented circular industries.

Furthermore, managing the exchange rate plays a vital role in controlling inflation within the domestic economy. A stable exchange rate can mitigate the impact of imported inflation, as fluctuations in the exchange rate directly affect the costs of imported goods and commodities. As Vietnam's economy continues to integrate into the global market, this aspect of monetary policy becomes increasingly critical to maintaining price stability. Stable prices encourage investment in circular practices and enhance consumer confidence in circular products.

In summary, Vietnam's monetary policy wields a multifaceted influence on the circular economy. While it provides stability in international trade and supports inflation control, the intricate interplay between money supply and interest rates requires careful consideration to ensure that policies effectively align with circular economy objectives. Achieving this balance is pivotal for fostering a sustainable and thriving circular economy in Vietnam.

4.2 Exogenous Monetary Policy Shocks

By estimating Eq. (11), an exogenous monetary policy shock (v^s) is determined, representing factors beyond the control of the SBV. The development of monetary policy shocks is illustrated in Fig. 1, part A. Due to the high volatility of v^s, a method proposed by Bernanke and Mihov (1998) and Fung and Yuan (2000) involved calculating the moving averages of v^s over six quarters. As a result, the v^s values fluctuate around zero, where a positive value signifies a monetary tightening policy shock and a negative value represents a monetary easing shock.

During the period from 2000 to 2005, the SBV implemented a monetary easing policy in line with the government's directives. Positive domestic reforms and various international economic-political events, such as the successful organization of the Asia Pacific Economic Cooperation (APEC) Summit and Vietnam's official entry as the 150th member of the WTO, fueled optimistic expectations and a new wave of investments in Vietnam. However, this surge in capital inflows led to currency appreciation, impacting the trade balance and causing a significant deficit. To counter the currency appreciation, the SBV intervened by buying foreign currency, resulting in an increase in the money supply. The monetary easing shock during this period was unexpected for the SBV, and the excessive money supply growth heightened inflationary pressures, leading to the implementation of a monetary tightening policy in 2008.

Following the inflationary period of 2018, the SBV adopted a revised approach to money supply, aiming to stimulate economic growth. However, the reforms in the Federal Reserve's interest rates had a significant impact on Vietnam's money and foreign currency markets, triggering unexpected tightening policy shocks. Consequently, there was a sharp increase in interest rates and a notable expansion of exchange rate fluctuations. Prior to 2016, the SBV's focus had been on averting currency depreciation pressures through its central exchange rate mechanism, which led to a greater emphasis on utilizing the credit market as a mean to support economic growth. However, unforeseen events such as the US-China Trade War in 2018 and the unprecedented COVID pandemic in 2020 introduced new tightening monetary shocks that the SBV could not anticipate.

Conducting a visual analysis of the effects of exogenous shocks on economic growth (measured through moving averages over four quarters), it becomes evident that an easing shock during the period 2004–2007 substantially boosted the economic growth rate, pushing it beyond 8% per year. The development monetary policy shocks and economic growth are illustrated in Figure 1, part B. However, the tightening shock, coupled with the global financial crisis in 2008, caused a sharp

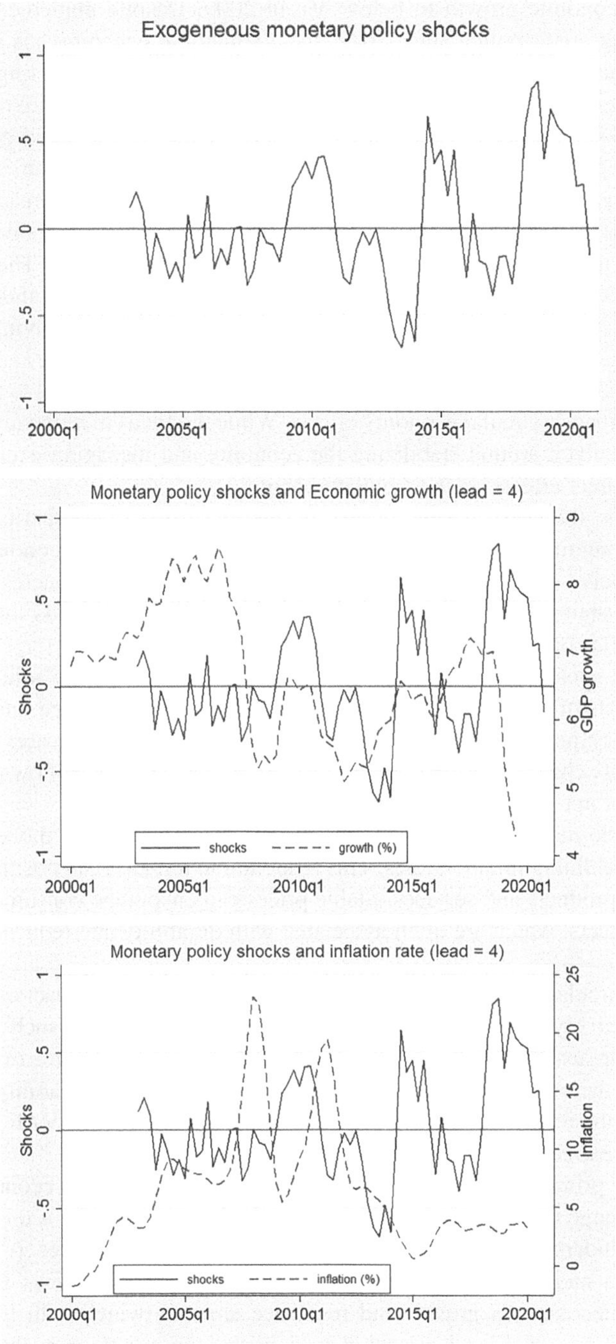

Fig. 1 Exogenous monetary policy shocks, inflation, and economic growth

decline in economic growth to below 5% in 2009. Despite implementing a cautiously easing credit policy since 2012, the Vietnamese economy has struggled to cope with the impact of exogenous shocks, hindering it from meeting its growth targets. In general, monetary easing shocks demonstrated a positive impact on economic growth, whereas tightening shocks restrained and reduced growth prospects. Correspondingly, inflationary pressures surged when facing an easing shock but gradually declined following a tightening shock, eventually reaching levels below 5% as shocks became less volatile. The development of monetary policy shocks and inflationary pressure is illustrated in Fig. 1, part C. These findings illustrate the complex relationship between monetary policy shocks and their repercussions on economic growth and inflation in Vietnam's ever-evolving economic landscape.

Amidst these monetary policy dynamics, it is essential to recognize the implications for Vietnam's circular economy efforts. While the focus of monetary policy has primarily revolved around stabilizing the economy and managing exchange rates, there are indirect effects on the circular economy.

Firstly, the monetary easing shocks observed during certain periods, such as 2004–2007, significantly boosted economic growth. A thriving economy can provide a conducive environment for circular economy initiatives by increasing investments in sustainable technologies and practices. These periods of economic expansion offer opportunities for circular businesses to flourish.

Secondly, exchange rate fluctuations caused by monetary policy adjustments can affect the costs of imported goods and materials. A stable exchange rate, as sought through policy interventions, can provide predictability for businesses engaged in circular supply chains. This predictability is crucial for the efficient flow of resources and materials in circular production processes.

Thirdly, the monetary tightening shocks that follow periods of monetary easing can lead to declining inflation rates. This reduction in inflation can positively impact the costs of products and services. Stable prices can encourage consumers to adopt circular products, which are often associated with durability and reduced long-term costs.

Finally, circular businesses in Vietnam need to navigate the impact of exogenous shocks effectively. The ability of these businesses to withstand such shocks and continue their sustainable operations is vital for the long-term success of the circular economy in the country. Monetary policy influences economic stability, which, in turn, affects the resilience of circular enterprises. For stability of SVAR model with and without exogenous shocks, see Appendix Figs. 3 and 4.

While the primary objectives of Vietnam's monetary policy are economic stability and exchange rate management, these policies have ripple effects on the circular economy. Understanding these dynamics and their consequences is crucial for policymakers and businesses seeking to advance circular practices in Vietnam. The circular economy's growth and resilience are intertwined with broader economic conditions, making it essential to consider these factors in the pursuit of sustainable and circular development.

4.3 Monetary Policy Stance Index

According to Bernanke and Mihov (1998), when the residual values of policy variables are substituted for the actual policy variables, a monetary policy stance index is made that includes both the endogenous and exogenous parts of monetary policy. The development of the monetary policy stance index is illustrated in Fig. 2, part A. To ensure consistency, the index is standardized by using a six-quarter moving average. A neutral stance is observed when the index centers around zero, while a positive index signals a tightening policy stance, and a negative index

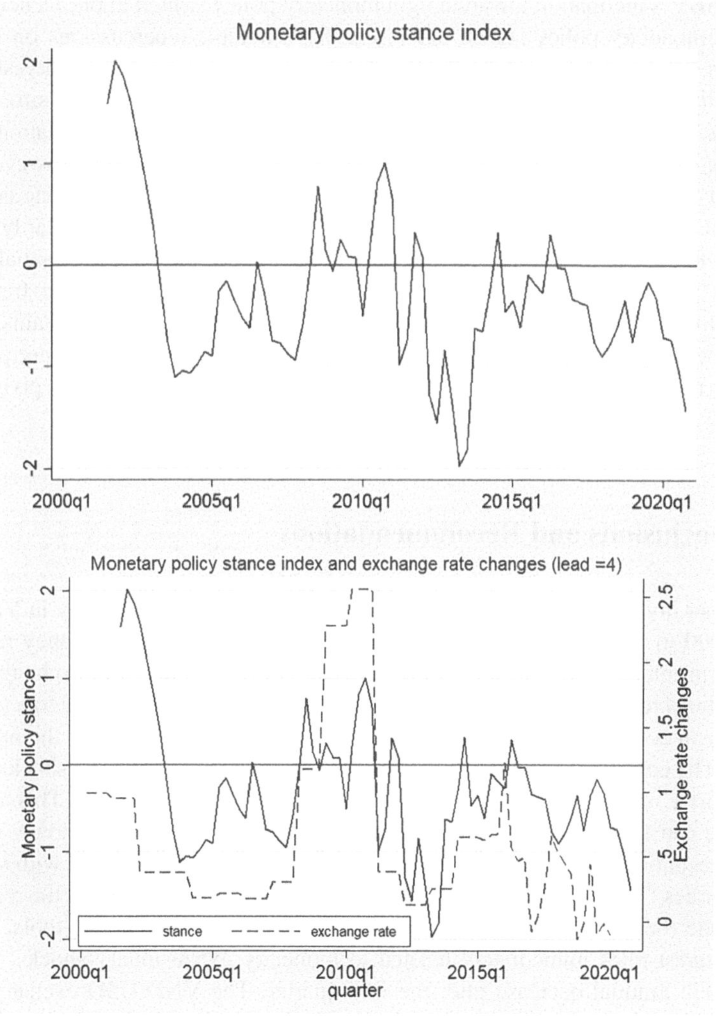

Fig. 2 The monetary policy stance index and exchange rate changes of Vietnam

reflects an easing policy stance. Figure 3 provides insight into the SBV's monetary approach, which predominantly centers on maintaining a tight monetary policy and employs policy tools to control the exchange rate. However, a noteworthy finding reveals a decline in the exchange rate fluctuation rate, which deviates from the associated pressures. This indicates a propensity for the monetary policy stance to align with fluctuations in the exchange rate. The relationship between monetary policy stance index and exchange rate changes is shown in Fig. 2, part B. The sudden increases in the exchange rate are attributed to domestic inflationary pressures and disparities between VND and USD interest rates. Since 2012, exchange rate changes have noticeably increased, largely as a result of pressures left over from previous monetary easing periods.

The SBV's inclination toward a tight monetary policy, which implies a neutral or positive monetary policy stance index, can have notable repercussions on investments in the circular economy. Such policies may result in elevated interest rates, potentially increasing the cost of financing for circular technologies and sustainable practices. Conversely, these policies can contribute to a decrease in inflation rates, lowering the expenses associated with circular products and services. However, it is essential to note that abrupt fluctuations in exchange rates can impact the costs of imported goods and materials, potentially affecting the viability of circular business models. In summary, the SBV's monetary policy stance exerts a substantial influence on various facets of Vietnam's circular economy, encompassing financing availability for circular initiatives, stability within circular supply chains, cost-efficiency of products, and the resilience of circular enterprises. Consequently, aligning monetary policy with the objectives of the circular economy is pivotal for fostering sustainable circular development in the country.

5 Conclusions and Recommendations

This paper investigates the transmission mechanism of monetary policy in Vietnam from 2000 to 2020, analyzing key policy variables, including the money reserve, short-term interest rates, lending interest rates, and the VND/USD exchange rate. The findings reveal that during this period, a monetary easing shock led to a gradual recovery in economic growth but failed to surpass the performance of the previous period. The consumer price index displayed a minimal response to the shocks in the first quarter, showing positive innovation only after four quarters. This limited response can be attributed to the challenge of transmitting the real money supply into the economy due to a decrease in short-term interest rates coupled with soaring lending rates. The exchange rate shock exerted significant pressure, yet the response magnitude ($\delta e = 0.006$) indicated strict control by the SBV's policy tools. Moreover, interest rates immediately reacted to monetary expansionary shocks, with a subsequent gradual decrease after the first quarter. The VND/USD exchange rate exhibited minimal response, experiencing a peak increase in the second quarter and a subsequent decline in the fourth quarter.

This study derives monetary policy recommendations to promote sustainable economic stability and growth in Vietnam, particularly focusing on advancing the circular economy.

1. Flexible exchange rate management: Instead of tightly controlling the exchange rate, a more adaptable approach is recommended. Allowing market forces to influence the exchange rate can better align it with economic fundamentals and external conditions. This shift toward flexibility can promote a balanced and stable economic environment. Concerns regarding increased debt burden can be mitigated by recognizing Vietnam's net borrower status, which reduces the impact of exchange rate risk.

2. Structural inflation management: Acknowledge that inflation risks primarily stem from domestic structural factors, such as supply-side constraints and imbalances, rather than solely from exchange rate policies. Addressing inflation necessitates structural reforms in the economy, including efforts to enhance productivity, foster competition, and optimize supply chain efficiency. By tackling the root causes of inflation through structural reforms, the economy can reduce its reliance on monetary policy as the sole instrument for price control.

3. Targeted money supply: When implementing a policy of easing the money supply, prioritize directing funds toward the real manufacturing sector. Channeling credit and liquidity into productive sectors can stimulate investment, technological advancement, and job creation. This approach fosters growth in the manufacturing sector, contributing to overall economic development and reducing dependence on nonproductive sectors, thereby lowering the risk of asset price bubbles. Policymakers should devise measures that incentivize banks to steer credit toward productive sectors and support initiatives promoting sustainable economic growth.

4. Stable inflation expectations: The SBV should prioritize maintaining stable inflation expectations among economic actors. Clear communication of the central bank's commitment to price stability and a transparent monetary policy framework can anchor inflation expectations. Consistency and predictability in monetary policy decisions can reduce uncertainty and enhance the effectiveness of policy measures. Employing forward guidance to communicate the central bank's future policy intentions can influence economic behavior and bolster the overall efficacy of monetary policy.

By adhering to these recommendations, Vietnamese policymakers can develop a more resilient and adaptive monetary policy approach, effectively addressing the unique challenges of the circular economy. A well-calibrated and targeted policy stance will contribute to sustained economic growth, stability, and resilience, effectively responding to external shocks and internal dynamics while advancing the circular economy's goals.

Declaration of Interest Statement No potential conflict of interest was reported by the authors.

Table 2 Results of unit root tests for the dataset

Variables	Dickey-Fuller test		Phillips-Perron test	
	Level	Difference 1	Level	Difference 1
LnRGDP	−3.552*	−8.514***	−1.399	−3.552*
LnRM	−2.537	−10.017***	−2.751	− 10.567***
LnCPI	−0.907	−5.220***	−0.483	−4.885***
LnEX	−1.274	−7.808***	−1.478	− 10.293***
LnVNI	−14.878***	−15.029***	−7.742***	−8.502***
R	−3.035	−8.199***	−2.388	−6.491***
TS	−2.978	−8.485***	−2.306	−6.380***
LnWCPI	−1.679	−6.427***	−1.543	−7.967***
FFR	−3.432*	−3.587*	−1.948	−4.649***

*Note: Values in the table are t-values. ***: $p < 0.001$; **: $p < 0.01$; *: $p < 0.05$*

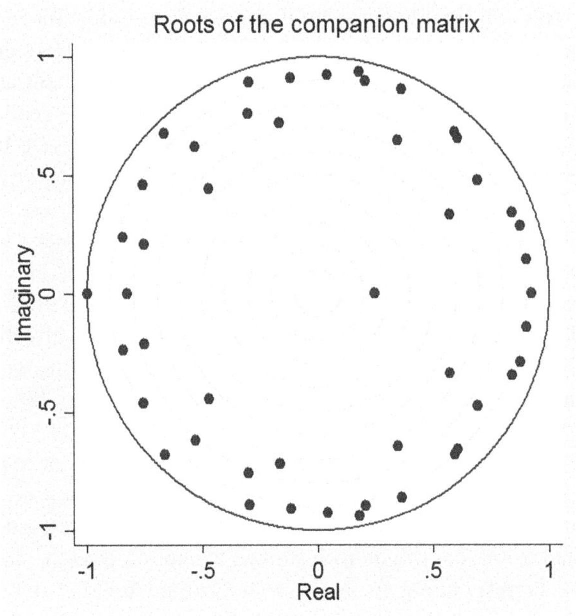

Fig. 3 Stability of SVAR model

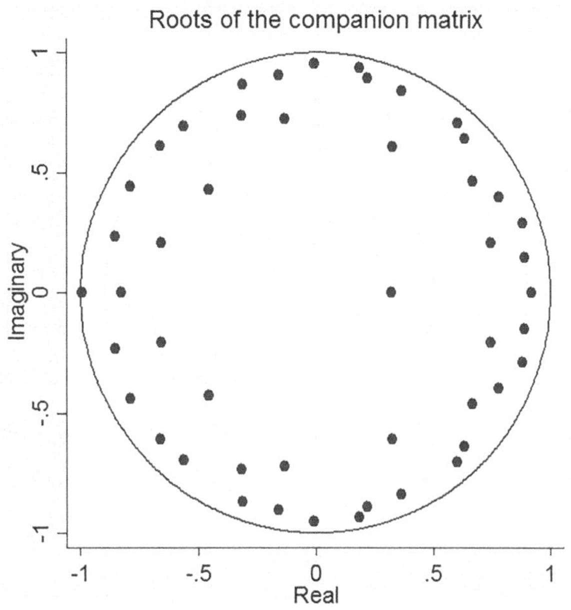

Fig. 4 Stability of SVAR model with exogenous shocks

Appendix

References

Amarasekara. (2008). The impact of monetary policy on economic growth and inflation in Sri Lanka. *Central Bank of Sri Lanka Staff Studies, 38*(1–2), 1–44. https://mpra.ub.uni-muenchen.de/64866/

Bernanke, B. S. (1986). Alternative explanations of the money - income correlation. *NBER Working Paper Series, 1842.*

Bernanke, B. S., & Blinder, A. S. (1992). The Federal Funds Rate and the channels of monetary transmission. *American Economic Review, 82*(4), 901–921. https://doi.org/10.2307/2117350

Bernanke, B. S., & Mihov, I. (1998). The liquidity effect and long-run neutrality. *Carnegie-Rochester Conference Series on Public Policy, 49*, 149–194. https://doi.org/10.1016/S0167-2231(99)00007-X

Christiano, L., & Eichenbaum, M. S. (1992). *Liquidity effects and the monetary transmission mechanism.* National Bureau of Economic Research Cambridge.

Christiano, L. J., & Ljungqvist, L. (1988). Money does granger-cause output in the bivariate money-output relation. *Journal of Monetary Economics, 22*(2), 217–235. https://doi.org/10.1016/0304-3932(88)90020-7

Dikau, S., Robins, N., & Volz, U. (2020). *A toolbox for sustainable crisis response measures for central banks and supervisors.*

Ferron, C., & Morel, R. (2014). *Smart unconventional monetary (SUMO) policies: giving impetus to green investment. Climate Report no. 46.*

Friedman, B. M. (1995). Does monetary policy affect real economy activity? Why do we still ask this question? *NBER Working Paper Series*, Art. 5212.

Fung, B. S. C., & Yuan, M. (2000). The stance of monetary policy, in money, monetary policy and Tranmission mechanism. In *Proceeding of conference held by the Bank of Canada* (pp. 233–262).

General Statistics Office of Vietnam. (2016). *Statistical yearbook of Vietnam 2015*. Statistical Publishing House.

Geweke, J. F., & Runkle, D. E. (1995). A fine time for monetary policy? *Federal Reserve Bank of Minneapolis, Quarterly Review, 19*(1), 18–31.

Husseina, S. A., & Hamdanb, A. A. (2020). The role of fiscal and monetary policy in financing and stimulating circular economy transition in Iraq. *AESTIMUM*, 125–145.

International Monetary Fund. (2014). *Vietnam 2014 article IV consultation-staff report; press release; and statement by the executive director for Vietnam*. https://www.imf.org/external/pubs/ft/scr/2014/cr14311.pdf

Keynes, J. M. (1913). *Indian currency and finance*. Macmillan and Company.

Law on the State Bank of Vietnam. (2010). *Pub. L. No. 46/2010/QH12*. National Political Publishing House.

Lucas, R. E., Jr. (1975). An equilibrium model of the business cycle. *Journal of Political Economy, 83*(6), 1113. https://doi.org/10.1086/260386

Murphy, R., & Hines, C. (2010). Green quantitative easing: Paying for the economy we need. In *Finance for the Future*. Norfolk. https://openaccess.city.ac.uk/id/eprint/16569/1/

Sims, C. A. (1980). Macroeconomics and reality. *Econometrica: Journal of the Econometric Society*, 1–48.

Sims, C. A. (1986). Are forecasting models usable for policy analysis? *Quarterly Review, 10*(Win), 2–16.

Strongin, S. (1995). The identification of monetary policy disturbances explaining the liquidity puzzle. *Journal of Monetary Economics, 35*(3), 463–497. https://doi.org/10.1016/0304-3932(95)01197-V

Barriers of Law Regulation to a Circular Economy in SMEs to Integrate in a Sustainable Development Strategy

Ngoan Võ

Abstract A circular economy is broadly defined as a regenerative industrial economy that captures vast amounts of resources for repeated use while ideally positively impacting the environment. It has been linked to significant economic benefits. The significant role of small and medium-sized enterprises in most economies is highlighted through an empirical analysis that explores the potential for implementing the circular economy in these enterprises, as well as the barriers to their transition toward a circular economy within a sustainable development strategy. Based on the described scenario of challenges and opportunities, this study aimed to make a legal analysis and show that companies are limited to complying with the law system in Vietnam. This study researches legal regulation systems to identify barriers and opportunities, which have been generally validated, and to raise awareness for taking action to encourage the development of SMEs in the circular economy.

Keywords Barriers · Circular economy · Regulation · SMEs

1 The Circular Economy Definition and Its Role

A circular economy usually has been described as a system aimed at eliminating waste and pollution while reducing the use of resources—it is a development strategy for a sustainable pattern for resource efficiency. The transition toward circular economy using the 3R system of reduce, reuse, and recycle influences the environment, the society, and the economy (John et al., 2023). If such a broad definition is used, circular economy and sustainable use of resources are the same. Different from what some people think, a circular economy is not a new way of waste recycling.

N. Võ (✉)
University of Economics Ho Chi Minh City, Vinh Long City, Vietnam
e-mail: ngoanvt@ueh.edu.vn

A. Heshmati, P. K. Nam (eds.), *Circular Economy and its Implementations in Southeast Asia*, Frontiers in South and Southeast Asian Development Research,
https://doi.org/10.1007/978-981-96-0539-2_5

A circular economy is broadly defined as a regenerative industrial economy that captures vast amounts of resources for repeated use while ideally, positively impacting the environment (MacArthur, 2013). A circular economy (CE) is a systemic approach to economic development designed to benefit businesses, society, and the environment. In contrast to the "take-make-waste" linear model, a circular economy is regenerative by design and aims to gradually decouple growth from the consumption of finite resources (MacArthur, 2013). In other way, a circular economy is an economy in which materials and products are efficiently and socially responsibly used within the ecological boundaries in order to enable future generations to share prosperity. It is a fundamentally different approach, a radical change of thinking and behavior. The circular economy has been gaining traction with business and government leaders alike. Their imagination is captured by the opportunity to gradually decouple economic growth from virgin resource inputs, encourage innovation, increase growth, and create more robust employment. If we transition to a circular economy, the impact will be felt across society, seen by specific benefits such as economic growth, material cost savings, job creation potential, and innovation.

A circular economy is an important pathway to realize the target of the sustainable use of resources, but it is not the same thing. Targets for a sustainable use of primary resources should further be concretized and quantified. The transition to a circular economy is a systemic change. For this transition, we will need new technologies, processes, business models, legal framework, and regulation.

The circular economy contributes to reducing waste in the environment. Developing the CE can help decrease approximately 48% of CO_2 emissions from transportation and construction activities by 2030 and up to 83% by 2050 (MacArthur, 2015). Furthermore, the CE aids in reducing the use of raw materials, minimizing the required landfill space for waste disposal, mitigating soil degradation, and reducing global waste processing costs.

The CE enables businesses to cut costs and avoid price volatility due to disruptions in global supply chains. Moreover, the CE necessitates businesses to invest in research and create new materials. It generates new business opportunities such as postconsumer services for collection and support of recycling products, marketing, and sales services to extend the lifespan of products and services for product regeneration. The aforementioned benefits demonstrate that applying the CE model yields advantages for the economy, society, environment, and enterprises. Implementing and promoting the development of the CE system can bring about global economic benefits of $4.5 trillion by 2030 (Rutqvist & Lacy, 2015) the Rutqvist and Lacy's report highlights the necessity of transitioning from a Resource Consumption Economy (RCE) to a CE. Circular economy contributes to economic growth of nations, measurable in terms of GDP (gross domestic product) and GNP (gross national product). By implication, circular economy provides opportunities for industries, businesses, governments, and society in the form of cost saving, job creation, innovation, and renewal (John et al., 2023; MacArthur, Growth within: A circular economy vision for a competitive Europr, 2015).

The sustainable development operates based on fundamental principles closely related to the efficient use of resources, recycling, circular production, processing, and resource utilization in an economical manner, minimizing and ultimately eliminating various types of environmentally polluting waste. There are three principles of sustainable development: First, conserve and enhance natural resources by controlling finite resources and balancing renewable resource flows; second, optimize resource usage by recycling products, product components, and materials at the highest level throughout both technical and biological life cycles; third, enhance system efficiency by identifying and excluding negative impacts from external sources (Nguyễn, 2022). After over years of innovation and development, Vietnam has achieved significant accomplishments. However, similar to many countries, Vietnam is facing numerous challenges such as natural resource depletion, environmental degradation and pollution, climate change, and more. Therefore, transitioning from a RCE to a circular economy model is crucial and prioritized in this new development phase, aiming of achieving sustainable development goals and fulfilling international commitments that Vietnam has engaged in.

2 Small and Medium-Sized Enterprises in Circular Economy

The vital role of small and medium-sized enterprises (SMEs) in most economies, which develops an empirical analysis that allows us to explore the potential for implementation of the circular economy in SMEs and the barriers and opportunities they may distinguish from it. SMEs are an important part of the private economic model and account for a large proportion of the Vietnamese business community. Every year, SMEs contribute 40% of GDP, 33% of industrial output value, and 30% of export value and attract nearly 60% of workers (Ministry of Planning and Investment, 2017). In 2019, the country had about 541,753 SMEs operating in the economic model, with a total registered capital of about 130 billion USD (The Central Economic Commission, 2019). This is also a difficult time as the world faces the COVID-19 pandemic and its effects. However, due to their small scale, the operations of these enterprises face many difficulties such as lack of capital to invest in modern machinery and technology to serve production and business and lack of experience in operating enterprises. The ability to compete and dominate the market even in the domestic market is still limited (Ministry of Planning and Investment, 2017).

In Vietnam, stakeholders' understanding of the circular economy philosophy and its benefits is still very limited (Nguyễn, 2014). SMEs today mainly produce based on a linear economic model, on the principle of exploiting resources from the natural environment as inputs for the economic system, through the production, consumption, and final processes discharged into the environment (Ninh, 2017). Therefore, the sustainability of businesses (economic, social, and environmental development) is the main issue for these businesses because they only focus on economic development and compliance with regulations. Desirable environmental and social goals according to global regulations and demands are becoming more and more difficult. Besides, the connection between SMEs in Vietnam is quite weak, and there are very few connections between small enterprises and even weaker large-scale enterprises.

During the COVID-19 epidemic, the economic situation changed. Before 2021, the pandemic exploded with major consequences and a new normal state, the circular economy model received even more attention in implementation and application. In the circular economy model such as effectively using output waste, saving input resources, and creating competitive products on the market, helping to bring production and business efficiency to businesses, many large businesses have also applied to adopt circular economy and have recently started encouraging their supply chains to also adopt circular economy. There is considerable heterogeneity among SMEs across different sectors—their responses, and capacities to take up a "green solution," are usually similar in terms of organizational and management regime. However, the adoption of circular economy in SMEs is very slow. Circular economy-related activities that SMEs are capable of implementing include waste minimization, energy planning, product and service design, renewable energy use, and reasonable water use.

The manager is usually also the owner of the company and thus has significant power on the strategic decisions of the firm. As such, some SME's managers may have a positive attitude toward green business, while others may not. This divergence of views toward "green business" has been attributed to a number of reasons in the available literature.

The vital role of SMEs in most economies, which develops an empirical analysis that allows us to explore the potential for implementation of the circular economy in SMEs and the barriers and opportunities they may distinguish from it. Numerous barriers can hamper the implementation of "circular" and "green" economy practices by SMEs that can originate, for example, from the SME enabling environment, such as culture and policymaking, from the market chain in which the SME operates. These barriers are further analyzed below on the basis of a literature review. Based on the described scenario of challenges and opportunities, this study aimed to address the following research about the main barriers and opportunities of the CE for SMEs?

Enhancing reuse and recycling of waste is only one of the transition tasks. Fostering the circular economy touches upon and creates a great variety of legal questions. Working, and hopefully answering some of these legal questions, will not bring about the circular economy and will not as such reduce our use of primary materials to a sustainable level. The transition of a circular economy urges that requires actions in not only the regulatory field but it also requires institutional changes, changes to accounting and financial instruments, cultural changes, technological innovation and knowledge development, and closer cooperation and transparency between all stakeholders. Legal research in this area must be integrated into interdisciplinary teams and projects. Whether and to what extent secondary materials can and will be used in the building of houses is a question which natural scientists, economists, political scientists, and lawyers have to discuss and solve together. How the use of insects that are grown on manure or other kinds of waste as food or for the production of food can be enhanced and how people can be moved to consume this kind of food are among others questions of engineering, behavioral science, and law. This study research about law regulation system to show the barriers and opportunities has been mainly validated in general, and it empirically validates them in the SMEs and in relation to the CE.

3 Research Methods

The research article primarily focuses on legal barriers to the sustainable development of small and medium-sized enterprises (SMEs) in the context of circular economy integration. Practical barriers arising from the implementation of legal regulations are studied based on real-life situations before and after the COVID-19 pandemic to obtain objective research results. To achieve good research outcomes, the article employs the following research methods:

Analysis and synthesis of theory: This method is used to synthesize information obtained regarding the circular economy and the contributions of SMEs to this economy and to analyze legal regulations both theoretically and practically.

Comparative method: This method is employed to draw objective conclusions based on comparing analyzed opinions and viewpoints.

Systematization method: Utilized to present the research content in a scientific and logical manner.

4 Barriers of Law Regulation

The writer based on the reality scenarios of SMEs in CE, case study, and review law regulation system, this study use observation to approach this topic.

Implementing the circular economy in Vietnam still faces many difficulties and challenges. The biggest challenge in applying the CE model is the cost of recovering value from waste. The circular economy is a closed-loop model that uses waste from one cycle as input for the next cycle. In Vietnam, the amount of waste is forecasted to double over the next 15 years. Vietnam's recycling rate is only below 10% of total waste. This is a small percentage compared to countries that have implemented the CE model. Nationwide, plastic waste and nylon bags currently account for approximately 8–12% of municipal solid waste. If about 10% of plastic waste is not reused and completely disposed of, the amount of plastic waste and nylon bags released into the environment would be approximately 2.5 million tons per year. The high waste rate creates difficulties in managing waste collection and resource recycling (Nguyễn, 2022).

4.1 Development of Circular Economy Policies

Numerous policies of the party and the state have been issued to promote sustainable economic development, such as Directive No.36/1998/CT-TW by the Politburo on enhancing environmental protection during the period of industrialization and modernization. This directive emphasizes supporting the application of clean technologies and resource-efficient practices. Additionally, the development of the CE model receives support from both the public and business. The Fourth Industrial Revolution and scientific advancements present significant opportunities to find solutions and enhance the efficient utilization of natural resources, environmental protection, and climate change mitigation while ensuring high economic efficiency.

In 2016, the government issued the National Action Program on Sustainable Production and Consumption. In 2017, the Scheme for the Development of Vietnam's Environment Industry until 2035 was approved by the Prime Minister to foster environmental industries aligning with the CE. In 2020, the Politburo issued Resolution No.55-NQ/TW on the direction of Vietnam's national energy development strategy until 2030, with a vision toward 2045. It emphasizes the priority of developing renewable energy, utilizing waste and residues in power plants, protecting the environment, and advancing the CE.

In 2021, the Resolution of the 13th National Party Congress affirmed the orientation of building a green economy, a CE, and an environment-friendly society. It also emphasized building roadmaps, mechanisms, policies, and laws to establish and operate the CE model. Furthermore, there are related laws and policies such as the Environment Protection Law, the Mineral Law, and laws related to marine and island environments.

In Vietnam, policies have emerged to encourage the application of CE in production processes within SMEs. The amended Environmental Protection Law in 2020 introduced Article 142 which refers to the CE concept, yet detailed guidelines or specific regulations for implementation are lacking. For instance, the use of recycled materials for more specific guidance could help relevant parties in applying these principles, especially in creating loops within production processes. Additionally, accessing social resources, credit capital, funds, and land for production and business still faces many difficulties, especially for small and medium-sized enterprises, leading to challenges in employees' social welfare due to the lack of support from public organizations.

The current economic system in Vietnam is leaning toward the needs of a linear economy. Enterprises prioritize market signals when making economic decisions, paying less attention to positive or negative external factors affecting society and the environment. CE production and business models are not widespread because most businesses still operate based on the logic of a linear economy, focusing on short-term value creation, while CE aims for long-term value creation.

Circular economy aims to change a linear economy into promoting sustainability of the economy while also engaging in sustainable environmental protection. The aim is to make a legal analysis and show that companies are limited to complying with the law system in Vietnam. There is still a lot to do in SMEs, as they are the companies that face the biggest challenges due to barriers of law regulation. Thus, this study raises the barriers to aware and have action to adjust the law regulation system to encourage SMEs' development in CE.

4.2 *The Law on Environmental Protection*

The Law on Environmental Protection No. 72/2020/QH14 (2020) is the most recent regulation that first ever had a direct mention of CE in Article 142. This Act goes into effect on January 1, 2022, although its Preliminary Assessment of Environmental Impacts went already into force on February 1, 2021. Companies generally participating in this initiative will adhere to Prime Minister's Decision No. 16/2015/QD-TTg, issued in May 2015, which outlines regulations for the collection and disposal of discarded products to mitigate electronic waste.

Furthermore, there is promise in Decision No. 687/QD-TTg, dated June 7, 2022, which endorses a program for the development of the circular economy in Vietnam. This program aims to serve as a catalyst for innovation and enhance labor productivity. It seeks to contribute to the advancement of green growth through economic restructuring, the adoption of a growth model emphasizing efficiency, fostering collaboration between businesses and economic sectors, enhancing the competitiveness and resilience of enterprises and supply chains against external shocks, and ultimately fostering economic prosperity and environmental sustainability.

The Deficiency and Constraints of the Law on Environmental Protection, from nearly 5 years of practical application, the Law on Environmental Protection has demonstrated that, alongside certain favourable outcomes, it also brings to light certain inadequacies and constraints. These encompass, among others:

The mechanism and policy for environmental protection lack alignment and synchronization with the framework of a market economy. Environmental tax and charges on the principle that "polluters must pay for environmental treatment, remediation, renovation, and restoration" and "beneficiaries of paid environmental value" are not distribute to promote the role of macroeconomic regulating tools, limiting activities causing pollution, environmental degradation, promoting socio-economic development in the direction of green growth (Bui & Kopytova, 2020). The clauses within the law have not established a conducive legal framework to incentivize sustainable production and consumption, foster the growth of environmental services, promote eco-friendly products, and facilitate the engagement of the community in certain environmental preservation endeavors.

The substance, allocation of duties and responsibilities, and decentralization of state-led environmental management exhibit shortcomings. It can be observed that the substance, allocation of duties and responsibilities, and decentralization pertaining to state-led environmental management still lack coherence, resulting in overlaps and contradictions. Furthermore, these aspects do not harmonize effectively with capacity enhancement and the unambiguous delineation of responsibilities, particularly in the spheres of urban and rural waste management and the rectification of environmental incidents.

The Law on Environmental Protection has revealed many inadequacies, lack of uniformity, and lack of regulations on new environmental issues:

Regarding the environmental impact assessment and environmental protection plan (Bui & Kopytova, 2020). The scope of environmental impact assessment is extensive, encompassing a multitude of elements that pose challenges in identification. There is a notable absence of explicit instructions for situations in which investors are unable to initiate or complete a project within a 24-month period. Moreover, specific regulations addressing modifications in project scale, capacity, technology, and the extent of content changes in repeated environmental impact assessments (EIAs) are lacking.

Concerning environmental protection in production, trade, and service activities, there is a deficiency in precise regulations regarding residential areas and the compilation of a catalog detailing the types of facilities and warehouses that require a secure separation from residential zones. The specific distances between these facilities and residential areas are not clearly defined. Furthermore, there is an absence of guidelines pertaining to the reuse of wastewater and the certification of environmental protection standards for facilities that handle daily-life and industrial solid waste treatment.

Regarding environmental monitor. The entities subject to monitoring lack clarity (for instance, whether an enterprise leasing a factory without wastewater discharge requires monitoring). There is incongruity in the frequency of observations, with disparities between environmental and water resource regulations. The efficacy of certifying eligibility for environmental monitoring services needs evaluation due to numerous inadequately equipped monitoring facilities. Additionally, the absence of a distinct category for automated monitoring compounds costs for businesses. There is lack of consistency in regulations on environmental monitoring objects specified in Clause 3, Article 121, and Clause 3, Article 123, of the Law on Environmental Protection (Bui & Kopytova, 2020).

In relation to pollution treatment, environmental restoration, and enhancement, there is a deficiency in precise regulations governing the obligations associated with environmental rehabilitation and restoration for insolvent enterprises. Specific duties of the Ministry of Natural Resources and Environment, along with state management bodies responsible for environmental matters, in responding to environmental incidents are not well-defined. Furthermore, criteria for identifying enterprises causing substantial environmental pollution are yet to be established.

4.3 Other Law-Related Challenges Facing SMEs

Regarding several other matters that still exhibit numerous deficiencies requiring examination, revision, and refinement, these include aspects like the overhaul of administrative procedures within the realm of environmental preservation; matters concerning economic instruments and resources for environmental protection; and delineations of duties for sociopolitical entities, organizations, and individuals in the realm of environmental conservation.

Besides, specific law-related challenges that SMEs might face when trying to adopt circular practices within their sustainable development strategies:

Waste classification and regulation: The prevailing waste regulations may lack comprehensive integration of circularity principles, potentially impeding the effective reuse, refurbishment, and recycling of materials. Rigid waste classification systems can pose a significant challenge for SMEs to navigate regulatory requirements.

Product design and extended producer responsibility: Legislative frameworks concerning product design and EPR may not sufficiently incentivize or mandate SMEs to incorporate design features facilitating ease of disassembly, repairability, and recyclability. Consequently, this deficiency may act as a deterrent for SMEs considering the adoption of circular business models.

Permitting and zoning regulations: The transition toward circular economy initiatives might require changes to manufacturing processes or the establishment of new facilities for recycling or refurbishment. However, intricate permitting and zoning regulations can serve as significant barriers, impeding or delaying SMEs from implementing such changes effectively.

Taxation and financial incentives: Current tax structures might not provide incentives for circular practices such as recycling, refurbishment, or remanufacturing. By recalibrating tax policies to recognize and reward circular initiatives, there exists the potential to motivate SMEs toward greater integration of circular principles. Traditional funding mechanisms might not be well-suited to support circular economy initiatives. Regulations governing financial products and funding opportunities could be adjusted to better cater to SMEs adopting circular practices.

Reporting and compliance burden: Reporting requirements related to circular practices could be onerous for SMEs. Streamlining reporting processes and aligning them with circular goals can help alleviate this burden. Absence of standardized guidelines for circular practices further compounds these challenges and underscores the need for their development and implementation. Additionally, difficulties and challenges arise from the demand for circular products and substitutes which are still small; a lack of qualified technical experts to implement this new economic model; and the current GDP measurement method does not prioritize social and environmental factors, failing to encourage value creation in both areas.

5 Discussion

The circular economy aims to transform the linear economy into one that promotes the sustainability of the economy while participating in the protection of the environment. The purpose is to analyze the legal aspects and highlight the constraints faced by small and medium-sized enterprises in complying with the legal system in Vietnam. There is still much to be done for SMEs as they face the greatest challenges due to legal regulations. Therefore, this study identifies barriers in perception and actions to adjust the legal framework to encourage the development of SMEs in the circular economy amid strategic sustainable development integration. Alongside certain advantages, the Environmental Protection Law also reveals certain shortcomings and limitations, which are described in the following sections.

5.1 Shortcomings and Limitations of Circular Economy

Absence of policies and provisions addressing emerging environmental protection issues Indeed, in recent years, there have been numerous instances of extensive pollution, significant environmental degradation, and widespread outbreaks of environmental crises due to emissions. Nonetheless, the existing legal framework lacks norms to establish mechanisms and criteria for identifying, categorizing, and stratifying investment projects based on their environmental risk levels. Moreover, specific control mechanisms aimed at high-risk projects that contribute to environmental pollution and incidents are currently absent. As a result, there is a compelling need to introduce supplementary regulations for screening and categorizing investment projects, accompanied by precise mechanisms to enhance oversight of these ventures.

The environmental administrative procedures exhibit fragmentation, lack of cohesion, and absence of integration. Within a single project, the investor is compelled to navigate numerous environmental administrative processes, seeking approvals from multiple environmental oversight entities as well as various governmental bodies. These include endorsements for environmental impact assessment reports, environmental protection plans, certificates of environmental project completion, permits for wastewater discharge into water sources, licenses for industrial emissions, and certifications for the importation of waste materials. Efforts to advance administrative reforms are imperative. This includes streamlining and interconnecting environmental administrative procedures outlined in the environmental protection laws, aligning with the guidelines set forth in Directive No. 30/CT-TTg, issued on October 30, 2018, by the Prime Minister. This directive aims to enhance the quality of administrative procedures across ministries, sectors, and local administrations (Bui & Kopytova, 2020).

The contents of environmental protection are regulated in many different laws (such as Law on Investment, Law on Public Investment, Law on Construction, Law on Water Resources, Law on Minerals, Law on Planning, etc.). However, between these laws, there are many interferences and inconsistencies, and some gaps have not been regulated, affecting the effectiveness of environmental protection state management. Provisions of the law only focus on the responsibility of the state, lack of appropriate mechanisms to promote the effective environmental protection participation of society, each people and enterprise. Results of reviewing and comparing the provisions of laws related to environmental protection also show that many provisions and provisions of the Law on Environmental Protection need to be revised and supplemented to: (1) Resolve conflicts between laws on environmental protection; (2) Updated to conform, consistent with the provisions of other laws enacted after such as Law on Investment, Law on Public Investment, Law on Construction, Law on Water Resources, Law on Minerals, Law on Planning...

5.2 Legal Conditions and Infrastructure for Circular Economy Development

Legal conditions and infrastructure for CE development are still lacking, with many shortcomings hindering the implementation of new business models. The circular economy requires upfront production and development strategies to use products for the longest possible time and plan to return materials for production later.

It's based on environmental challenges in Vietnam, the shortcomings and limitations of the Law on Environmental Protection which need to be amended and supplemented with suggestion the following recommendations are proposed:

In the context of formulating an environmental protection plan, it is imperative to meticulously revise the environmental conservation blueprint to ensure its seamless alignment with pertinent planning laws and regulations governing environmental preservation. This necessitates a thorough delineation of the foundational aspects of environmental safeguarding planning, which should encompass several key elements to enhance its effectiveness and legal compliance. Firstly, strategic measures for environmental protection must be intricately integrated within concurrent developmental phases. This involves a comprehensive assessment of potential environmental impacts at each stage of development, coupled with the formulation of proactive strategies aimed at mitigating adverse effects and promoting sustainable practices. Secondly, the inclusion of climate change projections is paramount throughout the corresponding development span. Given the increasingly evident effects of climate change, it is essential to incorporate long-term climate projections into the environmental protection plan to anticipate and adapt to future environmental challenges effectively. This may involve conducting thorough climate vulnerability assessments and integrating adaptive strategies to enhance resilience against climate-related risks. Furthermore, strict adherence to regulations governing environmental zones is essential to ensure legal compliance and effective environmental management. This entails compliance with zoning regulations that designate specific areas for different environmental purposes, such as conservation, protection, or development, and adhering to prescribed guidelines and restrictions within each designated zone.

In summary, the refinement of the environmental protection plan requires meticulous attention to detail and comprehensive consideration of legal requirements and environmental prerequisites. By incorporating strategic measures for environmental protection, integrating climate change projections, and adhering to regulations governing environmental zones, the revised plan can effectively safeguard the environment while ensuring legal compliance and sustainability.

Regarding environmental impact assessment (EIA). In order to enhance the effectiveness and quality of environmental impact assessments, suggest revisions and enhancements to the following aspects:

(1) Introduction of an additional preliminary EIA phase to ensure alignment with stipulations outlined in the Investment Law and Public Investment Law.
(2) Specification of EIA stakeholders to align with investment projects and construction investments as per the classification criteria delineated in the Public Investment Law.
(3) Augmentation of the procedure for evaluating EIA report contents concerning projects with potential adverse environmental implications. This reinforcement constitutes the foundation for enhancing EIA report quality, wherein the assessment's results hold project owners accountable for investing in waste treatment facilities.
(4) Recommendation to eliminate the necessity of an environmental protection plan for projects posing no risk of environmental pollution.
(5) Mandating the inclusion of EIA in the project preparation stage. Project proprietors may undertake the EIA process independently (or enlist consultants for support) and bear legal responsibility for the conclusive outcome of the assessment.

Regarding environmental administrative procedures. Currently there are seven types of environmental permits and discharging sewage into water sources. These permits are related and linked, so it is necessary to consider and integrate seven types of environmental permits and discharge wastewater into one environmental permit to reduce administrative procedures and reduce licensing costs for enterprises (Bui & Kopytova, 2020). It is strongly advised to assess the feasibility of consolidating the existing seven types of environmental permits associated with discharging wastewater into water sources into a single comprehensive environmental permit. Given the evident interrelatedness and mutual dependency among these permits, such integration holds substantial potential to streamline administrative processes significantly. By consolidating these permits into one, regulatory procedures can be simplified, thereby alleviating administrative burdens for enterprises. Moreover, this consolidation initiative has the added benefit of potentially reducing licensing costs for businesses, offering a pragmatic solution to enhance operational efficiency and regulatory compliance in environmental administrative procedures.

5.3 Waste Management and Environmental Protection

Regarding waste management. The waste management regulations of the Law on Environmental Protection must be amended and supplemented to the following contents:

(1) Issue the classification of waste, the responsibility to classify for recycling, reuse, and treatment to ensure compliance with environmental protection requirements, promoting the circulation and recycling of waste. This measure is crucial for ensuring compliance with environmental protection standards and fostering the circulation and recycling of waste materials, thus contributing to sustainable resource management.
(2) Issue regulations on management of radioactive waste transfer in accordance with the Law on Atomic Energy. Given the unique hazards posed by radioactive waste, comprehensive regulations are essential to mitigate risks to public health and the environment.
(3) Clearly define the responsibilities of waste owners, solid waste, and ordinary industrial solid waste treatment collectors based on "polluters pay" principle. It is paramount to clearly define the responsibilities of waste owners and collectors, particularly concerning solid waste and ordinary industrial solid waste. Adhering to the "polluters pay" principle, this clarification of responsibilities ensures that those generating waste bear the associated costs of its collection and treatment, incentivizing waste reduction and proper disposal practices.
(4) Clearly define the responsibilities of the provincial People's Committees, the investors, project owners, households, and small businesses and services in urban areas for collecting and treating wastewater up to standards.

It is crucial to delineate the specific responsibilities of various stakeholders, including provincial People's Committees, investors, project owners, households, and small businesses and services in urban areas, in the collection and treatment of wastewater to meet designated standards. By clearly defining these responsibilities, effective oversight and coordination can be facilitated, leading to improved wastewater management and environmental protection outcomes. Overall, these amendments and supplements to the waste management regulations are essential for enhancing environmental sustainability and ensuring the effective management of waste materials in line with national environmental objectives.

Regarding economic tools and resources for environmental protection. Therefore, it is advised to revise and enhance regulations to provide greater clarity regarding economic mechanisms, incentives pertaining to environmental conservation, and fees associated with environmental safeguarding. This entails the following specific measures:

(1) Introduce stipulations that promote the advancement of clean and renewable energy sources and the responsible and sustainable utilization of natural resources, ecological farming practices, sustainable patterns of production and consumption, eco-friendly lifestyles, and behaviors.
(2) Precisely delineate activities eligible for preferential treatment in the realm of environmental conservation, along with explicit support for endorsed environmental protection initiatives.
(3) Establish regulations pertaining to environmental levies, deposit-refund mechanisms, the transferability of emissions permits, compensation for ecosystem services, and the implementation of environmentally conscious credit practices.
(4) Guarantee the allocation of no less than 1% of the total state budget expenditure for environmental conservation expenditures.

5.4 Promotion of Circular Economy in SMEs

To overcome these law-related barriers and promote a circular economy in SMEs as part of a sustainable development strategy, policymakers could consider the following measures:

The provision of (nonfinancial) support by the (local) government represents a crucial facilitator identified in fostering a circular economy (CE). Governmental entities assume a pivotal role in advancing the requisite institutional frameworks conducive to CE development. This involves a multifaceted approach encompassing the reformation of extant laws, the enactment of novel regulations, the promotion of innovative environmental technologies, and the organization of public education initiatives. Corporate governance, when appropriately structured, can significantly augment the implementation of CE activities within organizations. For instance, the establishment of a dedicated and autonomous business unit oriented toward sustainable principles fundamentally alters the organizational structure, thereby exerting influence on behavioral patterns conducive to CE advancement.

A critical aspect of fostering a conducive regulatory environment for CE entails a comprehensive review and revision of existing regulations. Such an undertaking involves a meticulous analysis aimed at identifying regulatory barriers impeding circular practices. Subsequent amendments are then made to accommodate circular principles effectively. Additionally, the development of dedicated circular economy laws emerges as a crucial step in this process. Introducing new legal frameworks specifically tailored to incentivize and support circular economy initiatives, including provisions for incentives and guidelines geared toward small and medium-sized enterprises (SMEs), is imperative. Furthermore, initiatives aimed at educating SMEs about the legal ramifications associated with circular practices, along with providing training on compliance, intellectual property (IP) sharing, and other legal facets, are essential.

In parallel, the provision of clear guidance to SMEs regarding compliance with extant regulations while embracing circular practices is indispensable. Establishing support networks to aid SMEs in navigating the legal intricacies surrounding CE adoption is paramount. Engaging with SMEs, industry associations, environmental organizations, and legal experts to craft regulations that accurately reflect the unique realities and requirements of SMEs operating within a circular economy context is crucial for fostering a regulatory environment conducive to CE development.

6 Conclusion

The concept of the circular economy is increasingly acknowledged as indispensable for fostering and perpetuating sustainable growth across numerous nations worldwide. Within the context of Vietnam, although certain sustainable development initiatives have been initiated early on, their execution remains relatively limited in scope, characterized by a deficiency in systematic and all-encompassing integration. This research endeavors to present an impartial evaluation of the endeavors toward sustainable development within Vietnam while also proposing novel avenues for future research aimed at the cultivation of more nuanced, targeted, and efficacious strategies in fortifying the framework of sustainable development.

At its core, the circular economy seeks to transition from a linear economic model to one that prioritizes both economic sustainability and concerted efforts toward environmental preservation. Consequently, this study accentuates the imperative of addressing the barriers to awareness and action inherent in the current legal regulatory framework, particularly with regard to fostering the advancement of small and medium-sized enterprises within the circular economy paradigm. By addressing these impediments through legal reforms and aligning regulations to accommodate circular economy principles, SMEs stand to substantially enhance their capacity to incorporate circular practices into their sustainable development agendas. Consequently, this endeavor holds the promise of fostering a business environment that is not only more environmentally conscious but also economically resilient and conducive to long-term prosperity.

References

Bui, N. K., & Kopytova, A. (2020). Law on environmental protection to be revised for sustainable development in Vietnam. *E3S Web of Conferences, 164*, 11005.

John, I. B., Adekunle, S. A., & Aigbavboa, C. O. (2023). Adoption of circular economy by construction industry SMEs: Organisational growth transition study. *Sustainability, 15*(7).

MacArthur, E. (2013). Towards the circular economy: Economic and business rationale for an accelerated transition. *Journal of Industrial Ecology*.

MacArthur, E. (2015). Growth within: A circular economy vision for a competitive Europe. *Journal of Industrial Ecology*.

Ministry of Planning and Investment. (2017). *Vietnamese enterprises white book*. Thống kê.

Nguyễn, S.T. (2014). *Phát triển Doanh nghiệp nhỏ và vừa ở Việt Nam hiện nay*. Chính trị Quốc gia.

Nguyễn, L.T. (2022). Kinh tế tuần hoàn hướng tới phát triển bền vững. *Tạp chí Cộng sản*.

Ninh, T.T. (2017). *Hiệu quả kinh doanh của các Doanh nghiệp nhỏ và vừa ở Việt Nam hiện nay*. Chính trị Quốc gia.

Rutqvist, J., & Lacy, P. (2015). *Waste to wealth: The circular economy advantage*. Springer.

The Central Economic Commission. (2019). *2-year report on the implementation of the resolution of the 5th central committee of the 12th term on developing the private economy to become an important driving force of the socialist-oriented market economy*. Hà Nội.

International Experience on Financial Policy to Promote the Process of Collecting and Recycling Scrapped Vehicles and Lessons for Vietnam

Pham Hoang Ha and Pham Thi Thu Hong

Abstract Scrapped or end-of-life vehicle (ELV) disposal is expected to become a serious challenge for countries in the near future as the vehicle market in many countries continues to accelerate, leading to used vehicles and recycled parts to continue to expand. If ELV is not handled appropriately, serious social and environmental problems will arise and lead to damage to property. Therefore, to reduce waste to protect the environment, countries aim to encourage the reuse and recycle of ELV components by rapidly recovering ELVs. The country has been oriented to apply appropriate fiscal policy to support the recovery of discarded vehicles. The article studies the experiences of developed and developing countries in establishing fiscal policy to contribute to the recovery of waste vehicles and accelerate the process of recovering discarded vehicle from manufacturers and users, thereby providing lessons for Vietnam.

Keywords ELV · Recovery of waste vehicles · Disposal of scrapped vehicles · Recycling · Reuse · Taxes and fees

1 Introduction

End-of-life vehicle is defined as any vehicle that the owner ultimately intends to discard or does not wish to continue to keep (European Directive on ELV 2000/53/EC). An ELV is any vehicle that is disposed of, intended for disposal, or required to be disposed of, including obsolete vehicles, deregistered vehicles, damaged vehicles, and unused vehicles that have become obsolete (Nguyen, 2020). Vehicles are generally expected to last for about 15 years by manufacturers—those exceeding this age are classified as ELV (TNA et al., 2018). When the number of ELVs is increasing, it will become the main source of waste containing high levels of toxic substances from metals, plastics, rubber, and glass, causing serious problems for the

P. H. Ha · P. T. T. Hong (✉)
The National Institute for Finance, Ministry of Finance, Hanoi, Vietnam

A. Heshmati, P. K. Nam (eds.), *Circular Economy and its Implementations in Southeast Asia*, Frontiers in South and Southeast Asian Development Research,
https://doi.org/10.1007/978-981-96-0539-2_6

environment and people's health without a suitable recover and recycle system. Used cars, motorcycles, and their parts are important to speeding up the mechanization process in each country, but they also need to be serviced and recycled before being sold on the market to ensure safety for people and to protect the environment.

The ELV recovery and recycling process consists of four steps:

(1) The vehicle owner delivers the vehicle to the recalling facility (authorized dealers and shops).
(2) The company authorized to dispose of the ELV will carry out the dismantling of vehicle parts such as fuel, oil, tires, batteries or air-conditioning coolant that are taken to a landfill and other parts that can be disposed of. Other recyclables such as engines, suspensions, starters are sent to factories.
(3) ELV treatment plants perform decontamination and shredding of parts using a magnetic process to form a mixture of ferrous, non-ferrous and residues. The mixture is referred to as vehicle residues—Automotive Shedder Residue—ASR.
(4) Recycling plants receive ferrous mixtures and use eddy currents and density processes to separate aluminum, zinc, and copper (Sitinjak et al., 2022).

In the countries of the European Union (EU), in 2020, the number of ELVs is 5.4 million units, a decrease of 10.5% compared to 2019. Reuse and recycling rates for ELVs in the EU are 90.5% (much higher than the target set at 85% in the European Directive on ELV 2000/53/EC) (Eurostat, 2023), thanks to the establishment of a short-term financial support mechanism from the government for the EU. Disposal programs in 2009 encourage the dismantling of old traffic lights. In Japan, the current vehicle recall rate is about 80%—thanks to a financial sharing mechanism between the government and manufacturers, the remaining 20% consists of substances that cannot be separated or recycled (plastics, crystals) that are sent to landfills.

In other countries, the number of ELVs is increasing rapidly due to the trend of increasing vehicle ownership and decreasing vehicle life. The Australian Bureau of Statistics estimates that more than 500,000 ELVs are generated in Australia each year, and this recall will cost local authorities a lot of money (Levy, 2022). In 2022, Thailand has more than 5 million vehicles over 20 years old—it is forecasted that within the next 20 years, it will increase to 16 million units if there is no action to recall and recycle (Theparat, 2022). It is expected that by 2030, Malaysia will have 12 million vehicles in operation and have about 0.5 million ELV (Chong et al., 2022). Most countries such as Japan and Europe have good ELV waste management, but developing countries have not really done so due to the lack of effective fiscal policy to support implementation (Sitinjak et al., 2022).

There are many reasons that make this study significant and meaningful to Vietnam. Currently, the country has many outdated, expired, and unqualified means of transport still participating in traffic. This does not only threaten the safety of life for road users but also seriously affects the air quality of urban areas, health, and people's life. Vietnam currently has about 4.5 million cars and more than 60 million motorbikes in operation, and we forecast that the number of cars and motorbikes will continue to increase rapidly in the coming years. With the orientation that Vietnam will become a high-middle-income country by 2030 and strive to become a developed country by 2050, it is forecast that the percentage of people using motorbikes will tend to increase (from about 57% in 2019 to more than 71% in 2030) and the number of motorcycles in circulation increases from about 40 million units in 2019 to more than 63 million units in 2030 (Nguyen, 2019). For automobiles, the Strategy for the Development of Vietnam's automobile industry to 2025, with a vision to 2035, sets a target that the number of cars will reach 466,400 units by 2025 and 1,531,400 units by 2035. This poses many challenges in the recovery and recycling of waste vehicles soon.

In fact, Vietnam has issued a mechanism for implementing the recovery of ELVs from January 1, 2018, according to the Prime Minister's Decision 16/2015/QD-TTg dated May 22, 2015, but until now, the efficiency is not high due to the lack of financial policies to promote the recovery of ELVs by manufacturers as well as the voluntary participation of the people. Through the implementation of this decision, only Hanoi City piloted support up to four million VND to exchange old cars for new ones, but only four vehicles were recovered. The key thing is that people do not really think it is economically reasonable to dispose of their vehicles at the end of their life. Many people still consider expired cars as assets that can be used further or sold on the informal market. In addition, the informal purchasing facilities operate very effectively, making used car owners choose to sell to this market rather than bringing it to the manufacturer as it is today. In addition, according to the provisions of the Law on Environmental Protection 2020 (effective from January 1, 2022) and the government's Decree 08/2022/ND-CP dated January 10, 2022, from 2024 onward, enterprises produce and import six types of products such as batteries and accumulators, power electronics, tubes and tires, and lubricating oil; transportation and packaging teams will have to take responsibility for product recycling. As for vehicles, from 2027, manufacturers of cars and motorbikes will have to recover ELVs from people for recycling.

The document of the 13th National Congress of the Communist Party of Vietnam, the Socio-Economic Development Strategy for the period of 2021–2030, with a vision to 2045 has affirmed, "to encourage the development of a circular economy model to integrated and efficient use of the output of the production process". The National Strategy for Environmental Protection to 2030, with a vision to 2050, aims to prevent the increasing trend of environmental pollution by 2030, to solve urgent environmental problems, and step by step to improve and restore environmental quality. Through reviewing the experience of some countries around the world, tax, fee, and expenditure policies are important tools contributing to the implementation of sustainable development goals, including recovering the expired ELVs to protect the environment. Through the preliminary review, countries with many similarities with Vietnam such as India, Taiwan, Korea, China, etc. and developed countries such as the Netherlands, Germany, France, the UK, the Czech Republic, etc. have been successful in using financial policies for the recovery and recycle of ELVs.

Stemming from the status quo of ELVs in Vietnam, the orientation of the Vietnam Communist Party, and the negative effects on the environment brought by ELVs, this study has both scientific and practical meaning.

This study used qualitative method. We looked at documents, reports, and scientific studies of international organization and governments of countries through the Internet. Then, we apply policy analysis method for the content of the oversea policies on ELVs; we also use descriptive statistics method. Finally, we synthesize international experience into three sets of financial policies to apply to Vietnam, and we also provide our recommendation based on three sets of financial policies that we present.

2 Application of Financial Policies to Promote the ELV Processing

2.1 Policy on Taxes and Fees

Not only developed countries in Europe that have fiscal policies to promote the recovery of waste products but also countries in the ASEAN and Asia such as Indonesia, India, Taiwan, South Korea, Japan, etc. are also having effective policies. Most countries use tax policy, fees, and charges as a tool to regulate the behavior of car owners. These policies are all about the financial obligations associated with the possession of the vehicle, not the use of the vehicle. This means that the owner will have to fulfill financial obligations when the vehicle is registered even when the vehicle is not in use, so when the vehicle is not eligible for registration and circulation, the owner will automatically take the vehicle to a recycling facility instead of paying taxes, fees, and charges. Of course, there are other conditions required to encourage the vehicle owners to voluntarily bring their discarded vehicles to an official recycling factory. The condition is enforced by imposing penalties for the vehicle being ineligible but still on the road; have policies to support the exchange of old cars for new ones; public awareness of the need to dispose of expired vehicles; and licensed dismantling and recycling facilities and having an efficient, compliant, and convenient operating procedure for the owner of the disposal vehicle.

In the Netherlands, the ELV management system is considered one of the most developed in the EU. The distinguishing feature of vehicle registration in the Netherlands is the concept known as "possession." That means tax, inspection, and insurance obligations are related to the possession of the vehicle, not its use. For example, a vehicle must be in the registration system without interruption from initial stage to disposal (demolition or export). Motor vehicle tax (Dutch abbreviation mrb) is payable annually for use on public roads. Changes in ownership, the vehicle being exported, or converted to an ELV, and a certificate of destruction (CoD) being issued, the changes must be reported to the vehicle registry. As a result, Dutch authorities can easily track vehicles and ELVs. Vehicle owners will not have to pay motor vehicle tax, inspection, insurance, and tax obligations when they no longer own the vehicle, which is advisable when the vehicle is out of service life and ineligible. In circulation, the car owner will carry out procedures to dispose of the vehicle so as not to pay the above taxes and fees.

In Finland, when a vehicle is removed from the vehicle register or dismantled at an authorized ELV center, the owner receives a certificate of destruction. The insurance companies and the Finnish Transport Safety Authority (Trafi) were then notified of the destruction of the vehicle. As a result, auto insurance and vehicle taxes will be canceled, as a vehicle that enters the recycling system cannot be put back on the road. The vehicle tax, which is mandatory for all registered vehicles, includes base tax and engine-based tax, in the range of 100–500 euros.

In Ireland, vehicle owners pay an annual vehicle tax, which stops once a certificate of destruction is issued or the vehicle is sold or exported. For vehicles purchased before July 2008, the tax is based on the engine, while it varies according to CO2 emissions for vehicles purchased after July 2008, which has a tax rate of 120–2350 euros. However, if the vehicle is temporarily deregistered, the owner does not have to pay tax.

In Japan, consumers buying new and used vehicles are solely responsible for the ELV recycling fee at the time of their purchase. This fee is applied in addition to the cost of the vehicle purchase. Recycling fee is available ranging from 50 to 100 euros depending on the vehicle type and the number of airbags (the more difficult-to-recycle parts a vehicle has, the higher the recycling fee). This fee follows the car until the end of its life cycle and is managed by JARC (Wang et al., 2021), so when the owner sells the car to another owner, the two parties agree on the price of the car, knowing that when handing over the car, the service life has expired. To the recycling facilities, the end owner will receive this recycling fee along with a new car support based on the value of the old vehicle.

In Korea, disposal vehicle owners need to carry out the disposal at accredited facilities to be relieved of financial obligations to the vehicle owner such as vehicle tax, education tax, liability insurance, and inspection costs. Taxes, fees, and charges that car owners must pay in Korea are as follows:

The law stipulates the payment of taxes and vehicle usage fees, in which it is noteworthy that car owners must pay car tax (tax is calculated according to cylinder capacity, tax calculation coefficient, reduction factor, and tax based on the number of years of use of the vehicle); In addition, for vehicles not used for business, the vehicle education tax must be paid (equal to 30% of the automobile tax) (Pursuant to the Education Tax Law of Korea, 2016). The above two taxes must be paid every 6 months. The government imposes eight different taxes on passenger vehicles, which are levied based on the vehicle's price, insurance, and freight (CIF) plus an 8% tax rate. Three of the taxes are based on engine displacement. Korea's engine displacement tax is currently being applied so that a commensurate financial burden falls on vehicles with larger engines (over 2000 cc). All vehicles in Korea are required to have vehicle insurance. Types of car insurance in Korea include third party liability insurance: minimum legal requirements for drivers; comprehensive coverage: broader optional coverage covering a wide range of accident and financial coverages; and driver insurance: optional coverage for driver compensation and covers injury, fines, and consent in the event of an accident.

All vehicles in Korea must be checked regularly to ensure that they can stay on the road and minimize noise and pollution from exhaust emissions. These inspections are required by law and are overseen by the Korea Transportation Safety Administration (KOTSA). Vehicles are tested for safety, emissions, and noise standards. A new vehicle must be inspected prior to initial registration, and all vehicles must be inspected regularly. Tests are also required after any structural changes are made to the vehicle or its equipment. When disposing of the vehicle, the vehicle owner needs to take it to a licensed disposal facility and register the destruction with the local government office in their jurisdiction or face a fine (a fine of KRW 500,000). Accordingly, if the motor vehicle owner brings that motor vehicle to the recycler for processing, the recycler shall submit a dossier on behalf of the vehicle owner to cancel the registration of the recycled vehicle. This cancellation of registration will make it easier for vehicle owners to register new vehicles without having to pay taxes and fees for used cars to the state.

2.2 Budget Expenditure Policy

State budget mainly takes financial resources from the tax, fee, and charge to directly encourage owners to return the expired car and change it to a new car. Some experiences from other countries are taking effect:

In Denmark, in 2020, the country started an ELV collection program to provide payments to citizens who have entrusted ELV to an authorized treatment facility (ATF). Manufacturers in Denmark must provide owners with at least one collection point within a 25-km radius around cities with more than 20,000 inhabitants and within a 50-km radius around any location in Denmark. The economic incentive of the ELV program in Denmark is an ELV fund held by the Danish Internal Revenue Service (SKAT). This fund is funded from liability insurance paid by every vehicle owner every year. The current annual closing rate is 85 DKK (Danish krone, or about 12 euros). The costs for the ELV program in Denmark are covered from the ELV fund as mentioned. At the same time, each vehicle owner who ends up delivering ELV to ATF will receive a grant. Since the beginning of 2017, this amount can be up to 2200 DKK (about 300 euros) per vehicle (ADEME et al., 2019). Only light vehicles (cars carrying people up to nine seats) and light cargo vehicles (with a maximum total weight of 3500 kg) are eligible for assistance. Burned cars are also eligible for assistance. Support is paid to the last registered owner.

In Korea, regarding the financial mechanism for users, vehicle owners when disposing of vehicles need to carry out the treatment at accredited facilities to be relieved of financial obligations to vehicle owners such as vehicle taxes, education tax, liability insurance, and inspection costs. In addition to removing the financial burden of taxes, fees, and insurance, owners often also receive an amount for their ELV paid by the recyclers or producers when collecting and recycling ELVs based on the market price (Jang et al., 2022).

The Dutch extended the producer's responsibility (EPR) system for collecting and processing the ELV collection program.[1] It is organized collectively with a producer responsibility organization (PRO). (manufacturer's responsibility) only, Autorecycling Nederland (ARN).[2] ARN is also responsible for the management of most end-of-life vehicle batteries. To fund the organization's activities, ARN receives a recycling fee of 35 euros (2020) for each new vehicle purchased and a recycling fee of 5–180 euros for new batteries (depending on the weight of the battery). This recycling fee was mandatory until 2002 because RNA activities are supported by the government. Today, it has become a nonbinding agreement between the importer and the RNA. Now, 99% of vehicle importers pay a fee because ARN is responsible for enforcing the European ELV Directive and handling all new vehicles imported into the Netherlands.

In Japan, thanks to the mentioned 50–100 euros recycling fee, JARC can subsidize vehicle recycling operators if they recycle three components from end-of-life vehicles: gases fluorine (CFC/HFC/HCFC), airbags (two types), and shredded media (ASR). According to official data, the system is highly effective at ASR recovery reaching 97% and 85% for airbag recovery. In fact, as the ELV recovery and recycling process became heavily marketed, businesses competed for ELV resources. In such a scenario, when handing over their vehicle, the user not only recovers the recycling fee they paid but can also collect additional money paid by the recycling companies based on the condition of the vehicle convenience, and often much higher recycling fees.

2.3 *Financial Incentive Mechanism*

The EPR mechanism for ELV recycling has been in place in Taiwan since 1994. The 4-in-1 recycling program promoted by the Environment Agency (EPA) since January 1997 is a system that uses recycling fees, clearance, and handling fees collected from manufacturers and importers to establish a recycling fund, which is then used to subsidize the recycling system and expand the liability of these businesses.

---

[1] That is, the vehicle manufacturers or importers, instead of taking on the responsibility of recycling themselves, will authorize an intermediary organization (called PRO) to carry out the responsibility of vehicle recycling on their behalf.

[2] ARN, founded in 1995, collects and recycles waste from the vehicle sector. Use of ARN services is optional as it is a voluntary recycling program. In the Netherlands, 85% of the 220,000 ELVs are processed by 225 ELV centers located in the RNA network. Other ELV centers that handle the remaining 15% of ELV are of course subject to the same requirements as ARN members (with an RDW license and an environmental license issued by the environmental regulator), but they have their own processing system and is not controlled by RNA. These businesses, independent of the ARN, must also notify the government of their disposal rates for materials removed from the ELV (reporting required). ELV centers that are not part of the RNA network are not considered illegal operators.

Table 1 Recycling award in Taiwan 1999–2007 (unit: yuan/car)

Date	Car	Motorcycles
January 1, 1999–June 30, 1999	1800	400
July 1, 1999–December 31, 2000	3000	1000[a]
January 1, 2001–December 31, 2001	13,000**	3000[b]
January 1, 2002–December 31, 2003	3000	4000[c]
January 1, 2004–December 31, 2006	3000	1000 yen
January 1, 2007–June 30, 2007	3000[d]	1000[d]
From 01/01/2007	1000[d]	300[d]

Source: Chen et al. (2010)
[a]Notes: Vehicles over 7 years are subsidized 350 yuan/car
[b]Cars over 10 years old and motorbikes over 7 years old are subsidized at 10,000 yuan/car and 3000 yuan/motorcycle
[c]Motorcycles manufactured before 1997 are subsidized at 3000 yuan/motorcycle
[d]Only cars over 10 years old and motorbikes over 7 years old can receive the reward

Financial mechanisms adopted by the Taiwanese government to support and encourage ELV recycling include vehicle owner rewards, recycling fees, and recycling unit subsidies. Recycling fees, vehicle owner rewards, and subsidies for recycling operators are key to incentivizing stakeholders to engage in a sustainable recycling system. Recycling fees are paid by the importers and manufacturers to the Recycling Management Fund to be responsible for the recycling costs for the vehicles they manufacture or import. The recycling bonus is to incentivize people to bring their vehicles to operators for legal disposal. Audits and supervisory certifications help ensure that parties implementing the recycling process are compliant with regulations. The grants are used to assist the dismantling and recycling business to maximize recycling output.

2.3.1 Rewards for Car Owners

To incentivize vehicle owners to participate in the recycling system, vehicle owners are awarded cash rewards for delivering ELVs after unsubscribing. In addition to car owners, objects such as car dealers, service garages, and environmental police are also encouraged to participate in this program. If there is an unusable vehicle, the car dealer or valet service can bring it to a collection point and pay a fee. Environmental police often collect unusable cars on the roadside. If the owner is unknown, it will be treated as an ELV and then sold at auction.

From mid-1999 to the end of 2003, the Environmental Protection Agency (EPA) increased subsidies to owners who voluntarily surrendered their ELVs. The subsidies were funded with funding from the Air Pollution Control Fund (APCF) to reduce air pollution, gas and accelerate the replacement and recycling of old vehicles (see Table 1). The reward increases periodically until the end of 2003. The result of this incentive is an increasing number of ELV vehicles being recycled, while the subsidy is up to NT$10,000 per vehicle. However, due to the development of the recycling system and environmental concerns, the government reduced the reward in 2007. The adjustment of the reward value is still carried out when adjusting the policy (see also EPA, 2017).

2.3.2 Recycling Fee

The Recycling Fund Management Board (RFMB) provides a payment from the contributions of manufacturers and importers (called a recycling fee) to be able to assist whenever scrappers or recyclers need to invest additional investment to operate. The calculation of the recycling fee of the manufacturer and the importer is based on the number of vehicles produced in a given period and the corresponding number of imported vehicles. Table 2 shows the RFMB recycling fees received from manufacturers and importers.

Taiwan's vehicle recyclers now offer in-home collection, with tax breaks and subsidies available as other reasons for people to replace their old vehicles. When adding up the EPA's vehicle recycling subsidies, the merchandise tax reduction when replacing an old vehicle with a new vehicle, and the recovery value of the used car, a new car can save owners more than NT66,000 (about EUR 3000).

If a vehicle owner purchases and registers a new vehicle within 6 months of or after recycling the old vehicle, they will be eligible for the new vehicle merchandise tax deduction and the used vehicle recycling allowance. The tax breaks are NT50,000 and NT4,000, respectively, for cars and motorbikes. Vehicle recycling allowances are NT1,000 for cars over 10 years old and NT300 for motorbikes over 7 years old. For tax relief and recycling allowance, owners simply collect copies of the vehicle recycling manifest and vehicle title transfer form and apply to the Department of Finance for tax relief and with the EPA for reimbursement recycling allowance.

In addition to an additional NT1,500 to recycle motorcycles, the EPA and local environmental protection agencies also provide subsidies to purchase new electric

Table 2 Recycling fee paid by manufacturer and importer in Taiwan 1997–2010 (unit: yuan/car)

Date	Car	Motorcycles, mopeds
January 1997–October 1997	3000	700
November 1997–December 1997	2000	500
January 1998–June 1999	1000	250
July 1999–December 2000	983	264
January 2001–December 2004	643	96
January 2005–December 2006	965	144
January 2007–December 2007	2.050	683
January 2008–December 2009	3500	700
From January 2010	3800[a]; 2700[b]	800[a]; 650[b]

Source: Chen et al. (2010)
[a]Notes: The vehicle is designed in general
[b]Vehicles designed under the Green Mark Program

motorcycles and e-bikes. In addition, recyclers often pay owners several thousand NT dollars depending on the condition of the vehicle.

The EPA also points out that giving up or keeping but not using an old vehicle can cost owners almost NT16,000 annually in vehicle licensing taxes, fuel taxes, and penalties for not taking the vehicle away. If a vehicle is reported as occupying road space, the vehicle may be towed, and the vehicle's owner will receive notice and instructions to request its return in accordance with the Disposal Regulations. Abandoned vehicles occupy the road. If the vehicle is not returned, it will be disposed of, and the owner will not be able to claim a reduction of the goods tax or any subsidies, which is clearly not beneficial to the owner. The EPA is urging all used car owners to replace them with new cars and save money (at least NT66,000).

3 Financial Policy Lessons for Vietnam

ELV is considered a unique source of scrap and is beyond the control of the government of Vietnam. According to a study by Nguyen (2020), the rate of car ownership in households in Vietnam is increasing rapidly, estimated at 0.13 units by 2030, double in 2022, and there will be 190,000 discarded cars. However, more than half of ELVs (57% by weight) can be reused in Vietnam (consisting of engine parts, wheels, airbags, speakers, chassis, or mirrors), about one-third of ELVs (32%) can be recycled as a secondary material resource (mostly metal parts, some plastics and rubber), and the rest (11%) cannot be reused or Recycling is discharged to the dismantling area without control, causing many hazardous impacts on the environment and public health. While ELV is classified by the manufacturer as a group that needs to be recalled and handled under Decision 16/2015/QD-TTg, Vietnam does not have any official facility to handle ELV and lacks necessary regulations and support. Therefore, the problem of recycling ELVs is still limited in Vietnam, so it is necessary to have policies to accelerate the handling process for ELVs from owners and manufacturers.

Through studying international experience, several tax policies, fees, and subsidies from the state budget can be applied in Vietnam as follows:

Firstly, some places offer direct financial incentives by paying vehicle owners support when taking ELV to legal facilities (for example Denmark, Taiwan, etc.). This payment can be obtained from related taxes and fees or from the contributions of manufacturers and importers. However, determining a reasonable level of payment to compete with the informal sector should also be considered. On the other hand, taking into account the level of income, the price of cars sold in Vietnam is high compared to the rest of the world (about 2–3 times higher than in developed countries and about 1.5 times higher in countries in the region). Given the high value of a car as an asset, the amount of financial incentive to attract vehicle owners to dispose of the vehicles can be calculated. This is also associated with the need for a roadmap on how to reduce the cost of cars in Vietnam in the coming periods.

Second, direct financial incentive mechanisms should be applied at the same time as indirect mechanisms (in the Netherlands, Korea, etc.), that is, if the vehicle registration is deregistered legally, the owner of new vehicles is exempt from fixed financial liabilities with their vehicles such as taxes, vehicle fees, inspection, liability insurance, etc. This also requires systems to track the financial liability of the vehicle regularly throughout the registration and use of the vehicle. At the same time, amendments should be made to these financial liabilities in a way that discourages the use of second-hand vehicles, i.e., the longer the vehicle is in use, the higher the associated liability costs should be, rather than applying constant fixed costs regardless of the vehicle's useful life. This makes it much more expensive to own and use a used car than a new one and thereby is an incentive for car owners to take their old car to a disposal center to reduce the cost burden.

Third, the fact that car buyers must pay a prepaid recycling fee before buying a car (like the Japanese experience) is also a measure to consider raising awareness, responsibility and create financial resources to support for the ELV collecting and recycling system. This is a form of sharing responsibility for recycling and handling ELV from consumers with car manufacturers and importers. This fee is practically set by countries at a very low level (much lower than other taxes and fees when buying a car), so it doesn't change the price of the car much while it has the effect of raising the awareness and financial support for the ELV recycling management system.

Fourth, there is a need to study the reduction of taxes and fees when buying a new car with car buyers who have recycled their old car (according to experience in Taiwan) or car manufacturers and importers can apply programs vehicle rebates, gifts for new car buyers who have recycled ELV as a financial means to encourage ELV recycling.

4 Conclusion

Based on the basis of analysis and assessment of actual implementation of EPR for transport in some countries, lessons can be drawn for Vietnam as follows: (i) Regarding legal regulations: all countries promulgate regulations on vehicle ownership registration; revoke and cancel number plates; technical and environmental safety inspection for new and used cars, in which China and India are the two leading countries in determining the vehicle's service life. Most developed countries have enacted regulations governing ELV recycling except the USA and Australia. Regulations on facilities for collecting, dismantling, crushing, sorting, and treating waste are all approved for licensing if they ensure the regulations on area, equipment, finance, personnel, etc. (ii) The specific management model and the actual situation of ELV management in countries differ due to differences in their legal regulations. For example, in EU, China, and Taiwan, the responsibility for recycling rests with the manufacturer; but Japan and Korea are done by producers and consumers, and ELV collection system in Japan is done by collectors; in Korea, it is operated by a recycling facility or manufacturer or importer.

Accordingly, some suggestions are drawn for Vietnam: Basically, Vietnam's current regulations related to issues of registration, transfer, financial responsibility (taxes, fees, insurance), quality control, safety, and environmental requirements, etc. with motorized vehicles are also in accordance with international practices, but it is necessary to clarify the responsibility of vehicle owners to cancel vehicle registration and handle ELV of vehicle owners. In addition, it is necessary to develop regulations on the establishment and management of legal ELV collection, dismantling and recycling facilities to conduct ELV recycling in accordance with the law and ensure environmental protection.

Regarding the collection of ELV from the owner upon disposal, the regulations of some countries such as Denmark and the Netherlands require manufacturers and importers to set up ELV collection points within the appropriate range, while in other places, ELV treatment facilities will provide collection service or ELV collection point (Taiwan, Korea) to facilitate people to dispose of their ELV without charge. The fact that car buyers must pay a prepaid recycling fee before buying a car (like the Japanese experience) is also a measure to consider raising awareness and responsibility and create financial resources to support the system of collecting and recycling ELVs. In addition, it is also necessary to study the reduction of taxes and fees when buying a new car with car buyers who have recycled their old car (experience in Taiwan). Particularly for motorcycles, it is necessary to carry out periodic emission quality inspection as experienced in Taiwan.

References

ADEME (Eric LECOINTRE), Deloitte Développement Durable (Alexis LEMEILLET, Radia BENHALLAM, Antoine HENRY, Marie FILLION, Rafael BASCIANO), In Extenso Innovation Croissance (Victoire ESCALON, Beatriz BERTHOUX). (2019). *Final report: Global Overview of Incentive Schemes aiming to bring ELVs through Authorised Processing Channels*. 199 pages. Retrieved from https://copperalliance.org/wp-content/uploads/2021/08/global-overview-of-incentive-schemes-for-elvs-2019.pdf

Chen, K., Huang, S., & Lian, I. (2010). The development and prospects of the end-of-life vehicle recycling system in Taiwan. *Waste Management, 30*(8–9), 1661–1669. https://doi.org/10.1016/j.wasman.2010.03.015

Chong, J., Saman, M., & Ngadiman, A. (2022). End-of-life vehicles research development in Malaysia: A comprehensive review with the integrated conceptual model of innovative sustainable manufacturing elements. *Journal of Material Cycles and Waste Management, 25*(2), 698–716. https://doi.org/10.1007/s10163-022-01568-6

EPA. (2017). *Tax subsides available for replacement of old vehicles*. Retrieved from: https://www.epa.gov.tw/eng/F7AB26007B8FE8DF/475969e6-08fd-4552-a3fc-62901c18025b

Eurostat. (2023). *End-of-life vehicle statistics*. Retrieved from https://ec.europa.eu/eurostat/statistics-explained/index.php?title=End-of-life_vehicle_statistics#Number_of_end-of-life_vehicles

Jang, Y. C., Choi, K., Jeong, J. H., Kim, H., & Kim, J. G. (2022). Recycling and material-flow analysis of end-of-life vehicles towards resource circulation in South Korea. *Sustainability, 14*(3), 1270. https://doi.org/10.3390/su14031270

Levy, C. (2022). *Case studies*. Retrieved from https://circabc.europa.eu/ui/group/636f928d-2669-41d3-83db-093e90ca93a2/library/6e791763-4114-4143-bc11-e0e52072c605/details?download=true

Nguyen, D. (2019). *Chiến lược phát triển an toàn giao thông xe máy đến năm 2025, tầm nhìn 2030*. Retrieved from https://www.academia.edu/40844694/D%E1%BB%B0_B%C3%81O_T%C3%8CNH_H%C3%8CNH_GIAO_TH%C3%94NG_XE_M%C3%81Y_VI%E1%BB%86T_NAM_%C4%90%E1%BA%BEN_N%C4%82M_2025_T%E1%BA%A6M_NH%C3%8CN_2030_v%C3%A0_Chi%E1%BA%BFn_l%C6%B0%E1%BB%A3c_ph%C3%A1t_tri%E1%BB%83n_an_to%C3%A0n_giao_th%C3%B4ng_xe_m%C3%A1y

Nguyen, Q. (2020). End-of-life vehicles management and initial estimation of the material flow from ELV in Viet Nam. *Vietnam Journal of Science and Technology, 58*(5A), 138–149. https://doi.org/10.15625/2525-2518/58/5A/15223

Pursuant to the Education Tax Law. (1990, last amended 2016) of Korea, with the goal of ensuring revenue sources to increase spending on education.

Sitinjak, C., Ismail, R., Tahir, Z., Fajar, R., Simanullang, W. F., Bantu, E., Samuel, K., Rose, R. A. C., Yazid, M. R. M., & Harun, Z. (2022). Acceptance of ELV management: The role of social influence, knowledge, attitude, institutional trust, and health issues. *Sustainability, 14*(16), 10201. https://doi.org/10.3390/su141610201

Theparat, C. (2022). *Agencies urged to speed up ELV measures*. Retrieved from https://www.bangkokpost.com/business/2290194/agencies-urged-to-speed-up-elv-measures

TNA, R., MZ, M., Sharif, S., Simic, V., & AbdWahab, D. (2018). Development of a performance evaluation tool for end-of-life vehicle management system implementation using the analytic hierarchy process. *Waste Management & Research, 36*(12), 1210–1222. https://doi.org/10.1177/0734242X18790361

Wang, J., Sun, L., Fujii, M, Li, Y., Huang, Y., Murakami, S., et al. (2021). *Institutional, technology, and policies of end-of-life vehicle recycling industry and its indication on the circular economy comparative analysis between China and Japan*.

Electronic Banking Fraud Prevention in Vietnam in the Context of a Circular Economy

Ha Le Thuy

Abstract The article describes fraud in the banking sector, which is one of the risks arising in the circular economy, and highlights current methods of e-banking fraud through research. and learn about legal regulations domestically and internationally. In addition, the article also identifies the risks and challenges for preventing e-banking fraud in Vietnam today with a view to finding effective prevention solutions.

Keywords Fraud · Electronic banking · Digital transformation · Crime in the banking sector

1 Introduction

Digital transformation brings many benefits to the banking and finance industry because it helps commercial banks expand markets, reduce management and operating costs, and improve business rates and efficiency through products of digital conversion. One of the products of digital transformation is the birth and development of electronic banking. Electronic banking is the use of electronic and telecommunications networks to provide a variety of value-added products and services to bank customers (Steven, 2002). This is a good means of profit for financial institutions while providing convenient services with more flexibility and comfort than the traditional banking system. E-banking has made banking transactions easier by bringing services closer to customers, thereby improving banking service performance (Aduda & Kingoo, 2012).

In the context of unpredictable fluctuations in the global economy, along with the strong development of the 4.0 technology, the bank, in addition to facing traditional risks, also must face new challenges. Risks during the digital transformation process such as cyberattacks, Internet system crashes, customers' personal data being

H. Le Thuy (✉)
University of Law, Hue University, Hue, Vietnam
e-mail: hlthuy@hueuni.edu.vn

A. Heshmati, P. K. Nam (eds.), *Circular Economy and its Implementations in Southeast Asia*, Frontiers in South and Southeast Asian Development Research,
https://doi.org/10.1007/978-981-96-0539-2_7

hacked, etc. can cause extremely serious consequences for the company's finances and the reputation of a bank. An indispensable risk is bank fraud. The rise of online banking and convenience services has put them at risk of facing many risks and challenges due to attacks and the risk of data intrusion. This risk has reduced consumer confidence in the ability of financial institutions to protect them as well as anxiety about the safety of their money (Yazdanifard et al., 2011).

Digital transformation is considered a driving factor and provides solutions to the challenges that banks are facing in the digital era (Matt et al., 2015). Furthermore, according to Terrar, digital transformation is the process that an organization must go through when changing from outdated approaches to new ways of working and thinking by using digital, socialization, mobile, and new technology. But besides that, digital transformation of the banking industry also causes illegal acts to arise, exist, and develop—new forms of crime are born with sophisticated and complex tricks, taking advantage of the development of digital transformation in the banking system to carry out acts of appropriation of assets, causing impact, loss, and damage to the entire banking system.

Fraudulent acts arise partly from the risks themselves and are also the downsides of the circular economy. The circular economy is based on environmental, economic, and social aspects and aims to ensure sustainable development at every step, transforming products by creating a closed economy. In a circular economy like today, especially after the COVID pandemic, the conveniences of cyberspace have also made electronic fraud emerge as a challenge for the banking industry in particular and for customers and with countries in general. There, fraud perpetrators may enjoy luxury or waste that is contrary to the principles of the circular economy (Nagnonhou et al., 2023).

To explore the aspects of banking fraud and the downside of the circular economy of countries, including Vietnam, the article has set out research hypotheses including the following: What is the fraud in the banking sector? What methods of electronic banking fraud are there? Does Vietnam have any policies to prevent e-banking fraud? On that basis, the article also made suggestions to prevent electronic banking fraud.

2 Literature Review

2.1 Fraudulent Behavior in the Banking Sector

Fraud is a form of intentional deception to achieve the goals or needs of an individual or organization. In terms of criminology, fraud is understood as dishonesty, intentional deception, or intentional distortion of the truth, lying, intentional repetition of untrue events, etc. to gain an unfair advantage or disadvantage to another person (Binh, 2017). Fraud is synonymous with deception, conscious inaccuracy, against a person or organization to gain some advantage, while the other party is injured. The word fraud comes from the Latin *fraus*. Fraud is an intentional act of falsifying economic and financial information committed by one or more people on the board of directors, the board of directors, employees, or third parties falsifying records. Fraud is a type of intentional behavior that is often associated with self-interest (e.g., embezzlement, misappropriation of assets, falsification of information, falsification of records with the purpose of bringing benefits for an individual or a department).

From the perspective of criminal law, fraudulent acts can be considered a crime and punished, while from the perspective of civil law, those who commit fraudulent acts are forced to pay compensation for damages and harm caused to other subjects. There are different types of fraud, either specific fraud or general fraud. Specific fraud, as its name suggests, is the satisfaction of the legal hallmarks of a specific criminal offense, and otherwise it is called general fraud. In the laws of some other countries, fraud is also understood as intentional deception to secure an unfair or illegal advantage or to deprive the victim of legal rights.

Although there is no official definition of bank fraud, based on the concept of fraud mentioned above, it can be understood that bank fraud is the use of potentially illegal means to obtain money, property, or other property owned or held by a financial institution or to obtain money from a depositor by fraudulently representing a bank or other financial institution. Bank fraud is the intentional misrepresentation (usually requiring some technical expertise) to commit fraud (Benjamin & Samson, 2011). Types of fraud that financial institutions commonly encounter include sales fraud, purchase fraud, check cashing fraud, and ATM fraud, the taking of money or other assets from banks. Phishing is one of the mechanisms that fraudsters use to obtain customers' personal information leading to its use for fraudulent activities. Meanwhile, another view holds that fraud is a preconscious act by a person or group of people with the intention of changing the truth and/or reality for personal property gain. It involves the use of deception, sometimes even very sophisticated and calculated, such as forgery, falsification of documents and signature authorization, or an intentional act of theft (Adewumi, 1986).

Thus, it can be seen that fraud is understood to include forgery or errors and even fraud. While forgery is the falsification or alteration of documents and documents with the intention of deceiving, in some respects, fraud can be understood as a type of fraud. So, phishing is the tactics used by cybercriminals and scammers to get victims to reveal personal financial and other confidential information. From there, it can be understood that e-banking fraud is the act of using electronic means on online platforms to illegally carry out transactions in the banking sector by using deceitful tricks to usurp property or for other profit purposes. This behavior affects the banking system, affects online activities in the banking sector, and causes damage to the banks themselves as well as to customers who have placed their trust in them.

Electronic banking fraud is a form of online fraud that appears due to the development of the economy in general and e-commerce transactions in particular. Online fraud refers to activities that take place on the Internet or through e-commerce platforms, in which fraudsters use anonymity and Internet access to deceive customers and steal their personal or financial information. This is truly a potential threat as fraud criminals constantly change new tricks, behaviors, and tactics to evade detection and break safety and security measures. This makes businesses and management agencies face certain difficulties and challenges in dealing with the latest threats as well as taking effective preventive measures against risks from fraud and payment fraud online math. With the conveniences of electronic banking, consumers no longer have to spend time and money on traveling and performing financial transactions.

Although the ease and convenience of e-banking encourages individuals to make electronic transactions, however, using e-banking services and paying via credit card also expose them to cybercrime and can lead to their financial loss (Blut, 2016; Bressolles et al., 2014). The element of security and privacy in online transactions is an alarming issue that causes electronic fraud crimes to increase and take place almost everywhere and is one of the most serious threats (Campbell, 2019). E-banking fraud is the perpetrator taking advantage of the vulnerabilities of the online banking system or taking advantage of customers' indifference to issues of protecting their personal information to committing acts of fraud, appropriating assets, or causing damage to the banking system. Although banks have made efforts to prevent electronic fraud, large losses continue to be associated with fraud through electronic banking means, which indicates that financial fraud continues to occur, continues to pose a significant threat to banks.

Therefore, in order to protect electronic banking and minimize risks, especially minimizing fraudulent acts, it is necessary to access personal information security techniques to have a reliable security system along with awareness education for customers and related entities participating in and conducting banking transactions (Ahmad et al., 2021). These can also be seen as prevention tools to help consumers overcome risks in electronic banking.

2.2 Fraud Methods in Electronic Banking

Around the world, fraud methods in electronic banking are quite diverse, for example, stealing personal data to access a customer's account and then withdrawing money from it to another account. This method is familiar and popular in the early stages of electronic banking development. Another method that criminals are choosing to use is using websites with the wrong name or wrong address to commit fraud. As for phishing, there are many tactics used by cyber scammers, but the most important tactic is sending phishing emails to online banking customers by pretending to be a legitimate company or organization providing electronic services. Accordingly, a fake website has been designed by computer fraudsters to resemble legitimate financial institutions' websites to use for fraudulent activities and steal customers' financial information. Or the act of violating network security to spread viruses to users' computers to collect customer information to serve the criminals' evil purposes.

In foreign documents, fraud comes in many different forms and bank fraud is one of them. Bank fraud is also understood in a broad sense, including general fraud, corporate fraud, executive fraud, internal fraud, external fraud, computer fraud, leadership support fraud, recognized fraud judicial support. . .(Ojo, 2006). Or bank fraud also includes theft and embezzlement, falsification, forgery, suppression, fraudulent substitution, payment against unclear effects, illegal lending, informal borrowing, impersonation, Excessive invoicing, manipulation of documents, fictitious contracts, fictitious accounts, over/under valuation of assets, false declaration of cash shortages, fraudulent use of bank documents, falsifying status reports, abuse of suspended accounts, mail forwarding, blocking of check clearing, money laundering, computer fraud, false collections (Owho, 2005). This shows that fraud is understood in a broad sense, including many different forms, not just fraud.

In the abovementioned types, it can be seen that computer-assisted fraud is a specific form of electronic banking fraud. It is a process that involves all acts involving the use of computers through a fraudulent or dishonest manner or approach to deprive individuals of their rights or their property without having their consent. Here, computers are the main target of fraud. Computers are seen as a tool used by criminals to commit fraud through data entry, program manipulation, or operational manipulation such as tampering with computer files, manipulating input data, creating fictitious accounts, using another employee's password, removing fake applications, and using master password. According to Owolabi, e-banking also attracts many types of fraud such as stolen cards (physical card fraud), fraudulent applications, fraudulent manipulation of card data, spam or denial of service, fraudulent express access fraud, money transfer fraud, interbank clearing fraud, money laundering fraud, and mail fraud (Owolabi, 2010). Meanwhile, according to Dzomira, electronic fraud can be classified into two types: direct fraud and indirect fraud. Direct fraud includes credit/debit card fraud, employee embezzlement, and money laundering. Indirect fraud includes phishing, hacking, viruses, spam, advance fees, and malware.

Credit/debit card fraud and identity theft are two forms of electronic fraud that are often used interchangeably. It involves impersonation and identity theft (name, social insurance number (SIN), credit card number, or other identifying information) to commit fraudulent activities (Dzomira, 2014). In summary, in the world today, there are the following types of electronic banking fraud: phishing fraud, identity theft, credit card fraud, friendly fraud, and auction fraud. It is undeniable that fraud always goes hand in hand with online banking security because the increase in scale, nature, or severity of fraud means that bank security is weak or there is no truly high security system to prevent fraudulent acts from occurring. This has impacted customer psychology and reduced customer trust in the banking system. On the contrary, if fraud incidents occur infrequently, customers are maximally protected by strictly implemented security services at the bank, which will significantly minimize losses for customers, creating confidence for customers when using electronic banking services. However, even if the security service is good, if each customer does not personally conduct security operations when performing electronic banking transactions, criminals can still commit fraud to achieve criminal purposes.

Meanwhile, in Vietnam, current popular e-banking fraud methods include the following:

- First, impersonating government officials (police officers, tax authorities, courts) to scam customers and request money transfer
- Second, impersonating the victim to illegally create and link an e-wallet opened by the thief with the victim's bank account
- Third, impersonating a bank employee to exploit information on a debit card or bank account
- Fourth, fraud through mobile applications and fake links for customers to access and steal information to appropriate property
- Fifth, tricking victims into accessing and entering security information on fake websites
- Sixth, fake email from the bank notifying the bank that the Internet banking account is at risk and asking the customer to authenticate it
- Seventh, mistaken transfer method and instructions on refund procedures by sending a link asking customers to fill in personal information to appropriate money in the customer's account
- Eighth, stealing customer information transacting through unsafe websites to build fraud scenarios and appropriation from customer accounts

In Vietnam in recent years, there have also been e-banking fraud cases. For example, Sacombank had its credit card counterfeited in April 2017, and Dong A Bank also had a similar case in November 2018 (Phuong & Dien, 2021) or the case of impersonating a bank official to defraud and open a credit card to appropriate assets (Lam, 2022). Vietnam is a country with a transitional economy and is in the early stages of international and regional economic integration as well as in the process of moving toward a circular economy, so banking activities are greatly influenced by external factors. This makes banking activities increasingly complex and susceptible to negative impacts without careful management and slow policy innovation. When the necessary conditions for banking operations are not stable, the operations of the banking system fall into a state of instability, which can lead to systemic breakdown, seriously threatening national security and the country's socio-economic development. The consequence of the instability of the banking system is that violations in the banking sector have increased recently with many forms of violations, methods and tricks of crime being sophisticated, complex and diverse.

2.3 Vietnam's Legal Framework Regulates Electronic Banking Fraud

To prevent e-banking fraud due to the rapid increase in fraudulent acts in recent years, especially in the post-COVID context, governments have strengthened timely prevention measures. One of those measures is the promulgation of legal regulations to protect the legitimate rights and interests of individuals, organizations, businesses, and banks and to respond to fraud by the offender.

Within the international legal framework, countries around the world have also made efforts to jointly sign legal documents to prevent cybercrime in general, including crimes related to banking sector. This activity has also become a key component in the security strategies of many countries to ensure cyber safety and security.

In Europe, countries also have policies and strategies to fight high-tech crime. These organizations have joined together because they share common views and primary concerns about high-tech crime. The European Council of the European Union also established a committee of experts on crime in cyberspace in February 1997, which then went to a meeting on "mutual support in the fight against corruption crimes," money laundering, high-tech crime, sexual exploitation, organized crime" to promulgate the Council of Europe convention on cybercrime. It can be said that this is the most meaningful achievement demonstrating the international response to this type of crime. The goal of the convention is to ensure harmony in the criminal laws of countries in regulating crimes and at the same time provide regulations to promote and support each other in the process of handling crimes, especially criminals.

Afterward, in 2001 the Council of Europe issued the Convention on Cybercrime, in addition to the participation of members of the European Union, Canada, Japan, South Africa and the United States also participated. Negotiate the Convention as an observer. (Vatis, 2010). This is an international treaty aimed at harmonizing national laws on cybercrime, enhancing national capacity to investigate such crimes and enhancing cooperation in investigations.

The Cybercrime Convention requires parties to criminalize direct computer fraud (in which the perpetrator deceives the victim with the help of a computer system) and indirect computer fraud (in which the perpetrator directly deceives a computer system to perform some type of function or service that brings unjust financial benefit to the perpetrator or another person or causes financial loss to the victim), On the basis of the Cybercrime Convention, the European Council continues to issue a framework decision "On combating fraud and counterfeiting of non-cash payment instruments," including Article 2 stipulating "Crimes related to non-cash payment instruments", a convenient payment". One of the contents of this framework decision stipulates that, each member state takes appropriate measures to ensure that such actions are considered criminal offenses in cases where they are committed intentionally, at least in relation to credit cards, Euro check, other cards issued by financial institutions, travelers' cheques, Euro checks, other checks and bills (Kurylo et al., 2017).

Previously, Vietnam currently has a legal framework that regulates electronic banking activities quite promptly and covers actions related to electronic transaction, including: the Law on credit Institutions amended in 2017 and circulars of the State Bank regulating the scope of electronic banking activities, risk management of electronic banking activities; Civil Code 2015; Cyber Information Security Law 2015; Cyber Security Law 2018; Civil Procedure Code 2015 and many decrees and circulars guiding these laws. These documents also have regulations on information technology system safety standards, regulations on security and safety for Internet banking activities, regulations on customer identification and authentication, and regulations on customer identification and authentication. Specific regulations for each service activity are provided on electronic media. In addition, to handle violations in electronic transactions through the banking system, the State has also issued documents to regulate and handle administrative or criminal violations.

Over the past 5 years, along with the national digital transformation goal, Vietnam has issued a legal framework to regulate and promote the development of relationships in the banking sector, including Electronic Transaction Law No. 20/2023/QH15 (issued on June 22, 2023, replacing the 2005 Electronic Transaction Law) or most recently Decree 13/2023 dated April 17, 2023, on protection of personal data. Along with strategies and policies on digital transformation of the banking industry, Vietnam also issued legal documents to prevent and handle violations in the banking sector, including regulations to handle violations. To handle fraud in electronic banking transactions, Clause 7 of Article 28, of Decree No. 88/2019/ND-CP dated November 14, 2019, of the government stipulating sanctions for administrative violations in the field of currency and banking (amended) supplemented by Decree No. 143/2021/ND-CP dated December 31, 2021 stipulates a fine of 100–150 million VND for one of the following acts: Carrying out, organizing the implementation or facilitating use anonymous prepaid cards on the internet, mobile applications or withdraw cash; card switching, card transaction clearing, card transaction settlement not in accordance with legal regulations on bank card operations; Transfer the card acceptance device and QR Code to another party for use; accept card payments without a card payment contract; Illegal use of card acceptance devices and QR Codes.

From the perspective of criminal law, fraudulent acts in the banking sector through electronic transactions are also criminally handled according to the provisions of the Penal Code 2015. Accordingly, all fraudulent acts intended to appropriate property can be handled according to crimes of property infringement of an expropriative nature specified in Chapter 16 or handled for crimes in the field of information technology and telecommunications networks belonging to the group of crime of violating public order and public safety in Chapter 21. Currently, the characteristic of fraudulent behavior described in the Penal Code is fraud. According to the Penal Code 2015, fraudulent appropriation of property is a deceitful act to make others believe in order to carry out personal, illegal purposes. The motive of the person committing the fraudulent act is to gain appropriating the property of another person or causing another person to voluntarily hand over the property to the criminal subject due to misunderstanding of the truth. Vietnam has not yet stipulated a specific law on the crime of fraudulent appropriation of property in the banking sector. Therefore, all fraud cases in the banking sector are understood and handled by prosecution agencies as acts of fraudulent appropriation of property according to the provisions of Article 174 of the Penal Code 2015.

2.4 *Challenges Facing the Prevention of Electronic Banking Fraud in Vietnam*

Digital transformation brings many benefits to the banking and finance industry because it helps commercial banks expand markets, reduce management and operating costs, and improve business rates and efficiency through products digital transformation (Ngoc Bich, 2020). Therefore, the convenience of electronic banking is recognized. Nowadays, with just one click on your computer right at home, you can perform all banking transactions. E-banking carries out different types of activities ranging from online bill payments, fund transfers, and using ATMs to deposit and withdraw cash. This has completely changed human behavior in the way they spend money—even businesses use this method to transfer money to and from anywhere in the world to serve their business purposes (Yazdanifard et al., 2011).

Global digital transformation coupled with generational shifts and changes in consumer behavior correlate with today's growing fraud risk issues. In a study, the authors the authors pointed out the challenges facing electronic banking including: gaps in the legal framework or regulations that have not kept up with the advancement of science and technology; Fraud prevention capacity for digital banking transactions is not adequate; Information security issues have not accompanied the development of technology in the banking industry; and low awareness of banking risks among people in Vietnam (Hang & Mai, 2021). These things have created favorable conditions for fraud to arise, and fraudulent transactions and tricks are becoming increasingly sophisticated and difficult to detect.

In Vietnam, one of the reasons why e-banking fraud occurs is because Internet users in Vietnam today have very little knowledge about network security. They are willing to provide personal information to service providers or product sellers when conducting online transactions in general and electronic banking transactions in particular, easily bypassing security steps online when accessing the Internet. Fraud in the banking sector is not strange to the entire banking industry and to entities involved in the fight against crime. However, e-banking fraud is a new behavior, arising and developing with increasing nature and severity due to the need for strong digital transformation of the entire banking system. More importantly, electronic banking fraud also entails other crimes. These are cybercrimes also known as crimes in the field of information technology, corruption crimes, money laundering crimes, and other specific crimes whose criminal signs originate from acts of electronic fraud or related to electronic transactions arising in the banking sector. There are also risks and challenges to the fight against crime in general and crime in the banking sector in particular that managers need to consider.

The problem of fraud in the banking industry is not limited to any one economy, country, continent, or environment—it is a universal phenomenon. In principle, financial transactions often take place in the form of professional activities, with clear processes, regulations, and procedures, and are kept quite carefully. In fact, most economic crimes in the banking sector are associated with "fraud" behavior, and therefore, it is necessary to research and clearly understand the nature and mechanism of fraud in the banking sector is the basic for developing effective preventive measures.

In a study, it was shown that the challenges faced by banks mainly include technical weaknesses, lack of knowledge and awareness, lack of laws, and lack of resources to investigate cybercrime and strengthen necessary tools to combat electronic fraud (Owolabi, 2010). It is the lack of awareness among the population on how to maintain a minimum level of security regarding personal information or electronic assets. It is the lack of supervision by line managers or senior management of deviations from current electronic control procedures—it is collusion between staff and team people. There is also a lack of technical expertise, lack of fraud detection tools and technology, and lack of human resources to carry out investigations by law enforcement agencies themselves. Another problem is that the current business pressure to meet set targets has led to an increase in fraud cases. These are also true in the Vietnamese context. Another issue that we believe affects the emergence of e-banking fraud in our country today is that there are not enough laws and law enforcement to deal with cases of e-fraud.

2.5 The Penal Code 2015

From a criminal legislative perspective, the Penal Code 2015 only stipulates the crime of fraudulent appropriation of property in Article 174 or the crime of abuse of trust to appropriate property in Article 175 (both having fraudulent behavior is a characteristic sign) or the crime of property theft in Article 173 (using deceitful tricks) to criminally handle electronic banking fraud with the purpose of appropriating money, asset. However, we agree with the opinion of Owolabi, saying that fraud is a type of fraud, a prominent feature of fraud, but not all cases of fraud are fraud. Because the nature of fraud is the misrepresentation or concealment of material facts. Therefore, if we consider fraudulent behavior in electronic banking activities as fraudulent appropriation of property and prosecute this crime as a criminal offense, it will not meet the need to fight and prevent crime.

For the act of using the Internet to commit property appropriation, the Penal Code 2015 has regulations on the crime of illegally entering another person's computer network, telecommunications network, or electronic means in Article 289 and the crime of using computer networks, telecommunications networks, and electronic means to commit acts of property appropriation in Article 290. However, even if prosecuted for one of these crimes, the act of e-banking fraud is still not fully satisfying the signs constituting a crime. Because fraud in electronic banking activities damages the economic management order of the State, infringes on the operations of credit institutions and banks, and at the same time harms political rights and interests the rights of individuals and state organizations, especially property ownership rights. In terms of criminal tricks, criminals use deceitful tricks, openly or secretly through electronic means and the Internet of the banking system to commit crimes. In terms of purpose, the main purpose of the offender is to appropriate property—besides that, there may be other purposes, thus harming other objects protected by criminal law. Therefore, the current Penal Code does not stipulate a specific crime to handle this crime. That is a challenge in the fight against crime in the banking sector in general and electronic banking fraud in particular.

From the above challenges, to effectively prevent e-banking fraud, it is necessary to determine the responsibilities of the subjects participating in prevention. Electronic banking transaction activities include two main subjects: banks (electronic banking service providers) and customers (service users). The responsibility of the subjects to prevent fraud in electronic banking transactions depends largely on the terms of the service agreement—something the bank has control over and can adjust at will mine. To reduce fraud and promote the development of electronic banking services, the State needs to find ways to balance the rights and responsibilities between banks and customers.

Refer to the UK Banking Code issued by the Financial Conduct Authority and the UK Unfair Terms in Contracts with Consumers Regulations, both of which oblige banks to ensure fairness to our customers (Macdonald, 2002). According to this Code, if a bank does not comply with its internal processes and standard practices in the banking industry, when a fraudulent transaction occurs, it cannot be exempted from liability, regardless of any provisions, the service use conditions they set for customers. Or in the USA, forcing customers to bear the losses of unapproved transactions can be considered a failure of the payment system. Pushing responsibility onto customers will lead to more negative consequences than the benefits they gain. Therefore, the Electronic Funds Transfer Act (EFTA) of 1978 and the Truth in Lending Act (TILA) of 1968 stipulate that when dealing with fraudulent transactions, terms more favorable to customers are applied (Brandel & Olliff III, 1979).

Based on consulting and learning from legislative experiences of countries around the world, Vietnam needs to research and promulgate a code of conduct and handling of fraud in the banking sector to address the cases e-banking fraud. That regulation needs to consider the responsibilities of the parties when damage occurs due to fraud. Somehow, regulations must also ensure that customer interests come first instead of focusing only on bank operations and placing all responsibility on the customer, because, in the long run, customer trust in banks will decrease and customers will not choose electronic banking services from banks that do not stand on the customer's side to protect their rights.

Along with that is the improvement of laws on handling fraud in the banking sector. The banking sector is an important sector, banking sector crimes include many acts; therefore, in order to prevent this crime, it is necessary to define a specific group of crimes in the Penal Code that includes electronic banking fraud. In addition, the act of taking advantage of electronic means in the banking sector to commit other crimes such as money laundering, embezzlement, and property theft should also be considered a sign reflecting a significantly increased level of danger compared to the normal behavior of these types of crimes.

3 Conclusion

The problem of fraud in the banking industry is not limited to any one economy, country, continent, or environment—it is a universal phenomenon. In principle, financial transactions often take place in the form of professional activities, with clear processes, regulations, and procedures, and are kept quite carefully. The fact is that many crimes in the banking industry are associated with "fraudulent" behavior, and therefore, it is necessary to research and clearly understand the nature and mechanism of fraud in the banking sector is the basis for developing effective preventive measures.

Banks are the most important tool to help the government run the economy. They are places where large amounts of money and assets are concentrated, so they are also targets for criminals to carry out criminal activities to appropriate assets. Faced with the choice of "choose safety or choose convenience," it is necessary to determine the factors of convenience and safety that must be implemented in parallel (Shah, 2012). Besides, security is a factor that is always emphasized because it is considered the success of electronic banking. If a bank has security gaps, this can potentially lead to personal information being leaked, thereby causing financial losses, and can also create negative publicity in the media. Therefore, the article also concludes that Vietnam needs to have a mechanism to secure personal data during the operation of electronic banking.

In developing a mechanism, the government needs to soon promulgate a law on personal data protection. In addition, Vietnam needs to develop laws and sanctions to handle e-banking fraud. Only criminally handling the current crime of property appropriation fraud is not enough to deter and is not commensurate with the nature and level of danger to society of the act. In addition, Vietnam also needs to establish appropriate tools and policies to perfect the electronic banking system in the context of globalization, including perfecting criminal law to handle crimes of electronic fraud in the banking sector. That will create new driving forces for effective fraud prevention in the banking sector while eliminating risks arising in the process of promoting a circular economy in Vietnam in the coming time.

References

Adewumi, W. (1986). *Fraud in Banks*. Nigeria Institute of Bankers.

Aduda, J., & Kingoo, N. (2012). The relationship between electronic banking and financial performance among commercial banks in Kenya. *Journal of Finance and Investment Analysis, 1*(3), 99–118. p. 115.

Ahmad, I., Iqbal, S., Jamil, S., & Kamran, M. (2021). A systematic literature review of e-banking frauds: Current scenario and security techniques. *Linguistica Antverpiensia, 2*(2), 3509–3517.

Benjamin, O. A., & Samson, B. S. (2011). Effect of perceived inequality and perceived job insecurity on fraudulent intent of bank employees in Nigeria. *Europe's Journal of Psychology*, 99–111. http://search.ebscohost.com/login.aspx?direct=true&db=a9h&AN=60039658&site=ehost-live

Bich, N. (2020). *Digital transformation in the finance-banking industry: Need convenience and safety*, December 16, 2020, https://www.vietnamplus.vn/chuyen-doi-so-trong-nganh-tai-chinhngan-hang-can-thuan-tien-an-toan/682629.vnp, Accessed 10.05.22.

Binh, D. T. (2017). *Understanding criminological theory on preventing economic crimes and in the banking sector*, http://congan.travinh.gov.vn/ch26/275-Tim-hieu-ly-thuyet-toi-pham-learning-ve-room-ngua-toi-pham-Kinh-te-va-trong-linh-vuc-ngan-hang.html, Accessed 01.05.22.

Blut, M. (2016). E-service quality: Development of a hierarchical model. *Journal of Retailing, 92*(4), 500–517.

Brandel, R. E., & Olliff, E. A., III. (1979). The electronic fund transfer act: A primer. *Ohio State Law Journal, 40*, 531.

Bressolles, G., Durrieu, F., Senecal, S. J. J. O. R., & Services, C. (2014). A consumer typology based on e-service quality and e-satisfaction. *Journal of Retailing and Consumer Services, 21*(6), 889–896.

Campbell, C. C. (2019). Solutions for counteracting human deception in social engineering attacks. *Information Technology & People, 32*(5), 1130–1152.

Dzomira, S. (2014). Electronic fraud (cyber fraud) risk in the banking industry, Zimbabwe, risk governance and control. *Financial Markets and Institutions, 4*(2), 16–26.

Hang, L. T. T., & Mai, H. Q. (2021). Current status of digital banking development in Vietnam. *Financial Magazine*, (11/2021), 45–48.

Kurylo, M., Klochko, A., Timchenko, G., & Gulyk, A. (2017). Banking in Ukraine as an object of criminal and legal protection. *Banks & Bank Systems, 12*(4), 114–120.

Lam, P. (2022). *Impersonating a bank employee to trick him into opening a credit card*, July 11, 2022, https://dantri.com.vn/phap-luat/gia-danh-nhan-vien-ngan-hang-de-lua-mo-the-tin-dung-20220710172403895.html, Accessed 01.10.22.

Macdonald, E. (2002). Scope and fairness of the unfair terms in consumer contracts regulations: Director general of fair trading v first National Bank. *The Modern Law Review, 65*(5), 763–773.

Matt, C., Hess, T., & Benlian, A. (2015). Digital transformation strategies. *Business & Information Systems Engineering, 57*(5), 339–343.

Nagnonhou, S. R. B., Imoniana, J. O., Reginato, L., & Silva, W. L. (2023). Role of connectors in corporate fraud and corruptions in era of circular economy. *Social Sciences, 12*(3), 134.

Ojo, A. (2006). *Curbing fraud within the banking system: A banker's perspectives.* A.M. Continental Ltd.

Owho, O. (2005). *Bank frauds, causes and prevention: An empirical analysis.* ATT Books Ltd.

Owolabi, S. A. (2010). Fraud and fraudulent practices in Nigeria banking industry. *African Research Review, 4*, 3.

Phuong, N. V., & Dien, T. V. (2021). Cyber security risks and challenges in the banking sector in Vietnam. *HCMCOUJS-Economics and Business Administration, 16*(2), 30–44.

Shah, M. H. (2012). *Critical success factors in e-banking: A study of two UK*. Retail Banks.

Steven, A. (2002). *Information system: The information of e-business.* New Jersey: Natalie Anderson

Vatis, M. A. (2010, June). The council of Europe convention on cybercrime, In *Proceedings of a Workshop on Deterring Cyber Attacks: Informing Strategies and Developing Options for US Policy*. http://www.nap.edu/catalog/12997.html, 207–223.

Yazdanifard, R., et al. (2011). Electronic banking fraud: The need to enhance security and customer trust in online banking. *Advances in Information Sciences and Service Sciences, 3*(10), 505–509.

Part III
Circular economy transition and its implications for sustainable development

Closing the Loop: How Energy Efficiency Labeling Supports Circular Economy Goals

Mahirah Kamaludin, A. A. Azlina, Awang Noor Abd Ghani, and Zaiton Samdin

Abstract Energy labels serve as a crucial tool for educating consumers on the energy efficiency of appliances. These labels enable consumers to make wise decisions that save money and reduce their carbon footprint by providing precise information on energy consumption, while energy-efficient appliances promote sustainability by reducing energy consumption and greenhouse gas emissions and conserving resources. By conserving natural resources by minimizing resource extraction and waste generation, the circular economy promotes sustainable development for future generations. This study observes Malaysian public awareness and knowledge of energy-efficient labeling and its acceptance of energy-labeled appliances. The study collected primary data from 791 participants in selected states of Malaysia. About 74.97% of people are willing to pay more for energy-efficient appliances to support sustainability. The study found that age, employment status, and environmental awareness are important factors that drive the public to spend more on energy-labeled appliances. Energy labels provide information about energy consumption and the environmental impact of products, thus supporting the circular economy. Recognition of their relevance by policymakers, industry stakeholders, and the public is crucial. Besides, the energy labels play a vital role in accomplishing SDG 7, "Affordable and Clean Energy."

Keywords Circular economy · Energy label · Energy efficiency · Malaysia

---

M. Kamaludin (✉) · A. A. Azlina
Faculty of Business, Economics and Social Development, Universiti Malaysia Terengganu, Kuala Terengganu, Malaysia
e-mail: mahirah.k@umt.edu.my

A. N. A. Ghani
Malaysia Environmental Economic Association (MEEA), Kuala Lumpur, Selangor, Malaysia

Z. Samdin
Institute of Tropical Forestry and Forest Products, Universiti Putra Malaysia, Seri Kembangan, Selangor, Malaysia

A. Heshmati, P. K. Nam (eds.), *Circular Economy and its Implementations in Southeast Asia*, Frontiers in South and Southeast Asian Development Research,
https://doi.org/10.1007/978-981-96-0539-2_8

1 Introduction

The rise in carbon emissions and global warming have brought attention to their impact on economic activities. More emphasis has been placed on energy consumption and the availability of energy-efficient appliances in the market. Consumers are also considering their preferences when making purchases. Energy efficiency (EE) standards and labels serve as a straightforward and practical approach to assist consumers in their selection of household appliances. These standards are instrumental in transforming the electrical appliances market. Utilizing market pull and push strategies is a way of revolutionizing the market. By implementing energy performance standards, the "push" strategy eliminates the least efficient models, while labels "pull" customers toward higher efficiency appliances. This encourages manufacturers to produce more efficient models, creating a more productive market (Masjuki et al., 2000; Zhang et al., 2020). The EE labels are informative labels attached to manufactured products. They provide information about the product's energy performance, including energy use, efficiency, and cost. These labels help consumers make informed decisions about their purchases, as they can easily compare energy performance between products. Consumers can save money on their energy bills and reduce their environmental impact by choosing more efficient appliances. Energy labels also incentivize manufacturers to improve the efficiency of their products and promote energy efficiency as a selling point.

Besides, implementing energy efficiency labels catalyzes the advancement of sustainable practices and the realization of a circular economy. The goal of the circular economy model is to reduce waste, optimize resource utilization, and reduce environmental impacts. In this case, consumers can enjoy longer-lasting and durable options for high-quality materials and better designs by opting for energy-efficient products with guidance from energy labels. This is a crucial aspect of the circular economy, as it helps minimize the need for frequent replacements. By reducing energy consumption and waste production, the circular economy aims to reduce its environmental impact (Helena, 2022). When product labels communicate energy efficiency clearly, consumers can make informed decisions that benefit themselves and the environment. Next, consumers can reduce their energy consumption and, eventually, lower their energy bills by opting for energy-efficient products. This economic advantage motivates people to invest in eco-friendly products, which in turn helps decrease the demand for energy and minimize the environmental effects of resource extraction. Moreover, reducing energy consumption aligns with the circular economy's objective of conserving resources.

Circular economy strategies aim to prolong the lifespan of products. To achieve this objective, these strategies promote targeted actions, such as designing products that are easy to disassemble and reuse or employing energy labels that consider various factors such as material efficiency, durability, reparability, and recyclability. The principles of a circular economy have been adopted by the European Union for the management of waste. These principles place an emphasis on waste reduction, reuse, recycling, and recovery while minimizing disposal in landfills (Rada et al.,

2018). The European Commission and multiple eco-labels have advocated for including design for disassembly requirements in legislation or voluntary environmental instruments to encourage product lifespan extension and component harvesting for reuse (Paul et al., 2017).

Nations worldwide have implemented energy standards and labeling programs to encourage the adoption of energy-efficient products and safeguard the environment. These initiatives have led to notable advancements in the energy efficiency of various technologies. There are two types of energy efficiency (EE) promotional measures: mandatory and voluntary programs. For voluntary programs, manufacturers and suppliers can participate in a voluntary energy labeling program. This program enables them to add energy efficiency information to their products voluntarily. Companies can decide whether to label their products based on their preferences. For mandatory programs, certain products are legally required to display an energy label that shows their energy efficiency rating. These labeling programs are typically created through legislation or regulations, and manufacturers and suppliers must follow the requirements for the designated products.

There are two types of labeling systems: comparative labels and endorsement labels (Haq & Weiss, 2016; Mahlia et al., 2003). Comparative labels allow consumers to compare the EE of products of the same type in the market. In contrast, endorsement labels show that a product adheres to energy efficiency standards or other criteria established by a third-party group. However, the number of products with comparative labels is much smaller than those with endorsement labels. Endorsement labels are primarily voluntary and designed to cover only the top performers. Manufacturers must obtain attestation from a disinterested organization to post these labels. Energy Star is an example of an endorsement label. It was developed in the USA and adopted in the European Union (EU), Canada, and Japan.

1.1 Implementation of Energy Labels in the Southeast Asian Countries

Several Southeast Asian countries have taken positive steps toward implementing energy labels, providing consumers with vital information about a product's energy efficiency and performance. Energy labeling effectively manages energy consumption. Tharinya et al. (2017) demonstrated that energy consumption varies by region; in Greater Bangkok and central areas, it is due to more households, while in the north, northeast, and south, it is linked to income. The labeling programs implemented by Southeast Asian countries are displayed in Table 1. While Cambodia, Laos, Myanmar, and East Timor (Timor-Leste) have not yet implemented the energy label, they are actively working toward introducing the program in the future. Cambodia does not have a well-established energy labeling program compared to other countries. Still, there have been advancements in energy efficiency and environmental initiatives over time. Yau (2020) recommended including the

Table 1 Southeast Asian countries' implemented energy labels

Country	Program	Agency involved in energy labels	Obligation
Malaysia	Energy efficiency criteria for material and electrical equipment to qualify for the minimum energy performance standards star rating	Energy Commission (EC)	Mandatory
Indonesia	Minimum energy performance standards and energy efficiency labels	Ministry of Energy and Mineral Resources	Mandatory Voluntary
Thailand	Energy efficiency labeling no. 5	Electricity Generating Authority of Thailand (EGAT)	Voluntary
Singapore	Mandatory energy labeling scheme (MELS) and the minimum energy performance standards (MEPS)	National Environment Agency (NEA)	Mandatory
Vietnam	Energy efficiency labeling	Ministry of Industry and Trade (MoIT), Vietnam	Mandatory
Philippines	Philippine energy labeling program (PELP)	Department of Energy (DOE)	Mandatory
Brunei	Energy efficiency (Standards and Labelling) order 2021	Department of Energy, Brunei	Mandatory

Source: Based on sources of the official agency's website from the countries

minimum energy performance standards (MEPS) and energy labeling in energy efficiency strategies to enhance energy performance in Cambodia.

The MEPS sets out the minimum energy requirements for energy-related appliances to carry out their tasks. These standards are crucial in meeting government energy efficiency and climate change goals, as they promote energy performance improvements in consumers' appliances. Typically, a government energy efficiency agency is responsible for implementing MEPS.

1.2 Energy-Efficient (EE) Labeling in Malaysia

With the increase in income, the standard of living has been elevated, resulting in a surge in the utilization of household appliances. This, in turn, has led to a significant rise in household electricity consumption, causing a considerable increase in the release of CO_2 during production and consumption. Increasing electricity consumption negatively impacts the environment and cannot be ignored. Figure 1 illustrates the amount of CO_2 emissions in metric tons per capita in Malaysia that contribute to air pollution. The data indicates a significant rise in total emissions over the past 45 years. To address this issue, the Malaysian government has implemented the National Energy Efficient (EE) Policy to enhance the country's energy efficiency for sustainable development. As part of this policy, the government has introduced EE

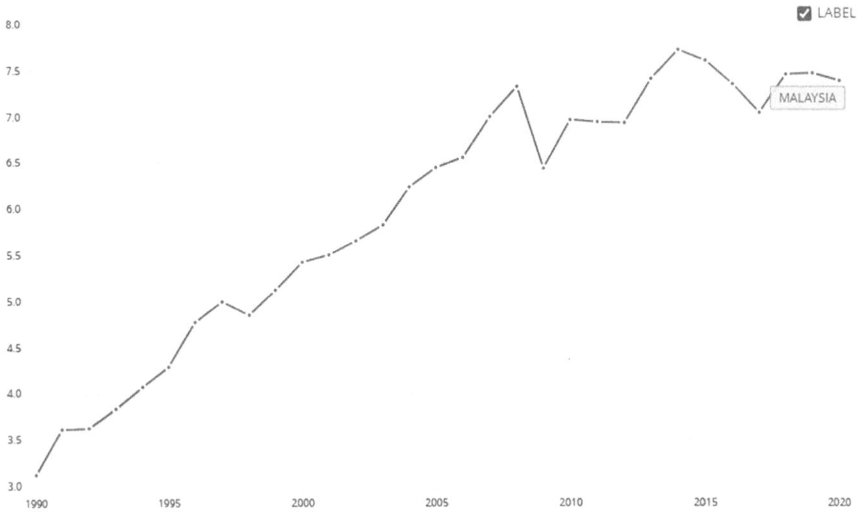

Fig. 1 CO_2 emission per capita (metric tons) in Malaysia from 1990 to 2020. *Source*: World Bank Data, 2023

labeling, encouraging producers to manufacture environmentally friendly or "green" products to support sustainable development. The EE labeling in Malaysia is under the strict supervision of the Energy Commission (EC) of Malaysia.

Implementing the labeling of appliances with an energy efficiency (EE) rating is a vital and efficient approach to educating customers about energy use. The Energy Efficiency Labelling Programme by the Energy Commission (EC) requires electrical appliances to be rated based on the STAR Index or efficacy value for lamp products, demonstrating compliance with energy efficiency standards. The initiative has been implemented to promote the use of appliances with maximum energy efficiency. In Malaysia, seven household appliances, such as televisions, refrigerators, fans, air conditioners, washing machines, microwave ovens, and rice cookers, must meet the minimum energy performance standards (MEPS) and achieve a two-star rating before selling. Similarly, lamps need to have a minimum efficacy value to be sold (Energy Commission of Malaysia, 2023). The EC mandates that manufacturers promote energy-efficient products in the market by issuing an EE label (Fig. 2), which is only granted after rigorous adherence to energy performance testing standards and requirements set by the EC for this rating.

Energy rating labels are a reliable and effective way of determining the energy consumption of electrical appliances (Fig. 2). Research by Mahlia et al. (2011) on the impact of energy guide labels on Malaysian consumers' purchasing decisions for household electrical appliances showed that many participants were unfamiliar with the label. However, it is worth noting that most of them understood and accepted its significance. The EE rating rates appliances from one to five stars, with five-star appliances being the most efficient and one-star appliances being the least efficient. This rating system is based on accredited laboratory test results and provides an

Fig. 2 Energy-efficient label applied in Malaysia. *Source*: Energy Commission of Malaysia (2023)

estimated total energy consumption in kilowatt-hours per appliance. Consumers can significantly reduce their electricity bills by choosing five-star-rated appliances, leading to higher disposable income and an improved quality of life. As per the Electricity Regulation 1994 (Amendments 2013) Regulation 101A (3), it is mandatory for appliances that comply with the efficient use of electricity requirements to bear an efficiency rating label.

The contribution of this study highlights the pivotal role of energy efficiency labeling in advancing the principles of the circular economy. By providing consumers with transparent information about the energy performance of products, these labels guide purchasing decisions toward more sustainable options, consequently incentivizing manufacturers to prioritize durability, reparability, and recyclability. This shift toward energy-efficient and circular products not only reduces resource consumption and waste generation but also stimulates innovation in product design and manufacturing processes. Moreover, energy efficiency labeling informs policy frameworks, enabling governments to set minimum standards for energy efficiency and integrate circularity into regulations. Overall, the study underscores how energy efficiency labeling serves as a catalyst for systemic change toward a circular economy, fostering collaboration between stakeholders and promoting a more sustainable and resource-efficient future.

The study aims to investigate (1) the awareness and knowledge among the Malaysian public regarding energy-efficient labeling and (2) factors that affect their acceptance of energy-labeled appliances. This label can influence purchasing decisions, promote energy conservation, and benefit the environment. The study is organized as follows: The introduction part emphasizes the significance of EE labeling, focusing on the developments in Southeast Asian nations and the introduction of the EE label in Malaysia. The Literature Review section critically examines prior research and existing literature regarding the significance of labels. The Research Methodology section elucidates the approach employed in the study. The Results section presents the survey findings and the outcomes derived from data analysis. Lastly, the Conclusion section provides a comprehensive summary of the entire study and its policy implications.

1.3 Literature Review

Energy labels offer numerous advantages to individuals, society, the nation, and the environment. Energy labels are necessary if we want to encourage energy efficiency and give consumers the ability to make educated choices about the amount of energy they use in their homes. They significantly impact consumers' purchasing behavior and encourage using energy-saving appliances (Morteza et al., 2020). Energy labels' effectiveness can vary depending on their design and implementation. The effectiveness of energy labels differs across countries, for instance, in Japan and Brazil (Guo et al., 2021; Mieko et al., 2020). Although some research has shown that energy labels can lead to energy conservation and positively influence consumer habits, other studies have raised doubts regarding the significance and actual impact of energy labels in the market (Cristian et al., 2020). Cristian et al. (2020) demonstrated that although the Brazil PBE Energy Efficiency Labelling Scheme has successfully enhanced energy efficiency, it has not substantially decreased energy consumption. When creating and implementing energy labeling systems, it is crucial to consider consumer cognition, perceived value, and trust in the energy labels (Sjors Van, 2019).

Besides, energy labels offer details about the energy efficiency of appliances, and this information can significantly impact the environment. An energy label is an effective method for promoting sustainable energy use and improving the environmental performance of products (Abhijit & Barry, 2003; Yang, 2020). Implementing energy labels is vital in the fight against climate change, allowing consumers to make informed decisions regarding their energy consumption and ultimately reducing their overall usage (Signe et al., 2015). The label empowers consumers to make informed choices in the market and encourages manufacturers to prioritize eco-friendly production practices. One effective way to convey a product's energy efficiency or sustainability is through labels or logos (Liuyang et al., 2011). Various energy labeling programs, overseen by both government and private organizations, exist to verify the energy efficiency of appliances and renewable electricity (Farshid et al., 2013). The government must support the programs to maintain their credibility, financial stability, and long-term sustainability (Liang, 2011). Yet, the effectiveness of energy labels relies entirely on the quality and quantity of information given, and how it is presented on the labels can influence consumers' decisions (Michalis et al., 2021).

Energy labels are critical in promoting energy conservation within the household appliance industry. Energy labels have been found to positively impact energy conservation efforts, which can be further enhanced by raising the reference point level (Mieko et al., 2020). Consumers can make informed decisions by offering valuable insights into the energy efficiency of appliances. However, implementing these labels varies from country to country, with varying levels of citizen awareness limiting their effectiveness. Despite their numerous benefits, the adoption rates of energy labels are declining over time (Anna et al., 2018). Research has revealed that consumers consider energy labels when deciding to purchase, and a home's energy efficiency can affect its price (Shigeharu & Yasuhiro, 2008).

2 Research Methodology

Between January and June 2021, we gathered data for a research project using a question-based interview format. The research sought to do a statistical survey with the objective of collecting data from a targeted group of people about their viewpoints, preferences, and understanding of energy-efficient labeling. To do this, we used systematic data gathering techniques, such as administering surveys and conducting in-person interviews. Respondents were given the opportunity to fill out parts of the questionnaire during the interview, leading to a mix of both approaches being included in the final instrument. The research aimed to improve the dependability and visibility of the data by collecting answers from 791 respondents in various locations of chosen states in Malaysia.

To investigate the variables that impact the public's acceptance of energy-labeled appliances, a logistic regression analysis was employed. Logistic regression analysis is frequently employed to explore the relationship between independent variables and binary or ordinal response probabilities. This study examined the dependent variable of "willingness to spend" on energy-labeled appliances. The logit model is utilized to express the willingness of the public to purchase and spend money on energy-labeled appliances, with "1" and "0" denoting "Yes" and "No," respectively, as in Eq. (1):

$$\text{Willingness to spend} = \alpha + \beta s + \varepsilon \tag{1}$$

where this study's survey comprised three sections, primarily examining respondents' viewpoints toward obtaining energy-efficient labels. In certain sections of the questionnaire, specific questions were repeated to gauge the comprehension of respondents on particular concerns. In Section A, the study explores the reasons that motivate respondents to spend more on energy labels, while Section B investigates the respondents' perception and knowledge of EE labeling on electrical appliances. Finally, Section C is designed to elicit information on respondents' sociodemographic profiles. These survey questions assessed respondents' knowledge and awareness of energy labels, environmental issues, and climate change. As such, it should be noted that not all questions were incorporated in this study. The survey questionnaire implemented in this study solely consisted of the questions recorded in Table 2.

Respondents were surveyed about their knowledge and attitudes toward energy labels, environmental issues, and climate change.

3 Results

Table 3 contains some critical characteristics of the respondents who took part in the survey. The survey participants consisted mostly of females, with an average age range of 21–30 years and a secondary level education. Significantly, the private sector had the largest presence in the sample population, with a notable 31.23% of respondents belonging to this category.

3.1 Respondents' Willingness to Spend on Energy-Efficient Appliances

Figure 3 displays the respondents' responses to spending extra money on energy-efficient appliances. As per the results, 101 (12.77%) respondents indicated a negative response, while 690 (87.23%) respondents expressed a positive response.

Table 2 Questions used in the study

Question no.	Questions	Response
Section A	*Questions on willingness to spend on energy-efficient appliances*	
SBQ6	Would you be willing to spend extra money to purchase an electrical appliance that is more energy-efficient?	Yes; No
SBQ7	Reasons why some individuals are willing to spend	
SBQ8	Reasons why some individuals are not willing to spend	
Section B	*Questions on the importance of energy efficiency labels in promoting sustainable energy use*	
SCQ9(1)	Purchasing electrical appliances that have an energy-efficient label can save you money on your electricity bill	Ordinal Likert scale 1–5
SCQ9(2)	Purchasing electrical appliances that have an energy-efficient label shows that the buyer is aware of the environment	
SCQ9(3)	Purchasing electrical appliances with an energy-efficient label as an effort to help preserve the environment for future generations	
SCQ9(4)	When I purchase an electrical appliance with an energy-efficient label, I can trust that it is both energy-efficient and of good quality	
SCQ9(8)	Purchasing electrical appliances that have an energy-efficient label can reduce energy waste	
SC10(1)	Discussing the issue of climate change is of immense significance	
SC10(3)	Science and technology can deal with the problem of climate change	
SC10(4)	I have been hearing a lot in the news about climate change	
SC10(5)	Carbon emissions are the primary cause of climate change on a global scale	
SC10(9)	Energy efficiency can have a positive impact on the environment, including the issue of carbon emissions	
Section C	*Questions about the sociodemographic profile of respondents*	
	Gender, education, employment status, household income, residential areas	

Miranda et al. (2022) demonstrated that energy-efficient appliances may come with a higher initial cost than traditional appliances. Purchasing energy-efficient appliances is a wise investment, despite the higher upfront costs. These appliances are designed to consume less energy, leading to substantial savings on utility bills over an extended period.

Table 4 displays why the respondents prefer to spend more on energy-efficient appliances. Energy-efficient appliances often cost more due to their energy-saving features. Most respondents (74.97%) were willing to spend more on energy-efficient appliances due to their commitment to support environmental sustainability. Energy-efficient appliances are designed to consume less energy while providing the same or improved functionality, reducing their carbon footprint by emitting fewer greenhouse gases. Renate (2017) demonstrated the critical role played by enhancing the

Table 3 Summary table of sociodemographic characteristics

Variables	Number	%
Gender		
Men	361	45.64
Women	430	54.36
Age group (years)		
≤ 20	11	1.39
21–30	256	32.36
31–40	225	28.45
41–50	152	19.22
51–60	120	15.17
≥60	27	3.41
Education level		
No formal education	9	1.14
Primary	23	2.91
Secondary	342	43.24
High school certificate/diploma	209	26.42
Bachelor's degree	188	23.77
Master/doctorate degree	20	2.53
Employment status		
Government staff	183	23.14
Private staff	247	31.23
Businessman	232	29.33
Farmer/fisherman	11	1.39
Unemployed	51	6.45
Others	67	8.47
Household income (MYR = Malaysian ringgit)		
≤2000	430	54.36
2001–4000	279	35.27
4001–6000	40	5.06
6001–8000	17	2.15
8001–10,000	16	2.02
≥ 10,000	9	1.14
Residential area		
Urban	365	46.14
Rural	426	53.86
Total	791	

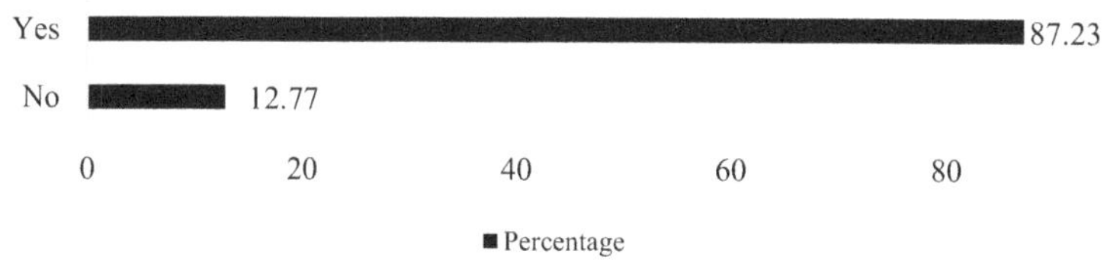

Fig. 3 Respondents' responses on their willingness to spend on energy-efficient appliances

Table 4 Reasons respondents agree to spend more

Response "Yes"	Number	Percentage
A sense of responsibility toward environmental sustainability	593	74.97
A sense of responsibility toward future generations	198	25.03
Total	791	

Table 5 Reasons respondents disagree to spend more

Response "No"	Number	Percentage
I cannot afford to purchase energy-efficient appliances	601	75.98
The government should bear energy conservation costs	190	24.02
Total	791	

energy efficiency of electric appliances procured by households, reducing energy demand, and promoting sustainability.

Respondents were also inquiring about their rationale for refraining from getting energy-efficient appliances. According to the findings in Table 5, it can be inferred that a significant proportion (75.98%) of the respondents expressed their inability to afford energy-efficient appliances due to their relatively higher cost than conventional appliances. This highlights the cost factor as a major deterrent in adopting energy-efficient appliances, which has important implications for sustainable development initiatives. Approximately 24.02% of the respondents believed that energy conservation costs should fall on the government rather than the public. This implies that the respondents rejected the idea of transferring the cost of energy conservation to the public and advocated for a government-led approach to tackle this issue.

3.2 Perceptions and Knowledge of Energy-Efficient Labeling

This study observes Malaysians' perception, knowledge, and awareness of energy-efficient labeling, particularly regarding energy-efficient appliances, energy consumption, and environmental issues. Section B is divided into two parts: one focused on energy-efficient appliances and the other on environmental issues.

Based on Fig. 4, 45.13% of respondents believed purchasing electrical appliances with energy labels reduced energy waste, while 3.41% totally disagreed. Energy-efficient appliances are designed to maintain the same level of performance while consuming less energy. These appliances include LED light bulbs, smart thermostats, ENERGY STAR-rated appliances, and high-efficiency heating, ventilation, and air-conditioning (HVAC) systems. Using energy-efficient appliances can significantly reduce energy consumption, lowering energy bills and reducing carbon footprints. Thus, adopting energy-efficient appliances is a significant step toward achieving sustainable development. Based on Fig. 4, it can be inferred that a significant majority of the respondents, approximately 42.7%, totally agreed that

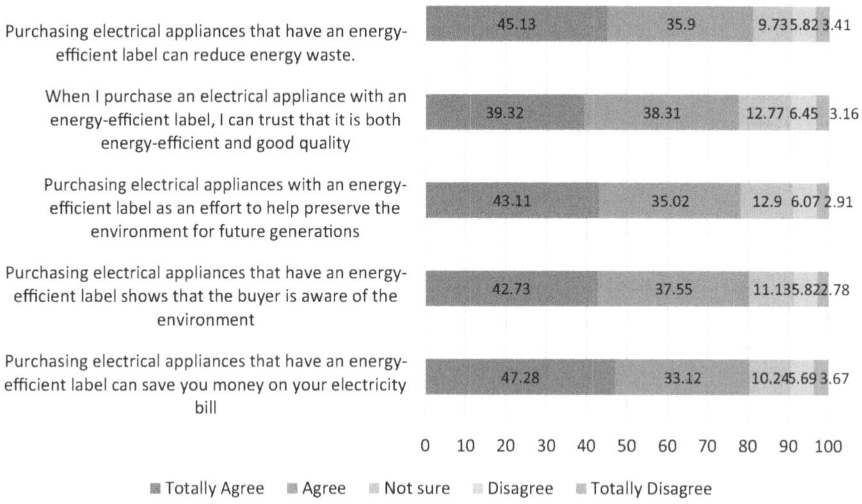

Fig. 4 Percentage distribution of knowledge and perceptions on energy-efficient labels

the discussion of climate change-related issues is of paramount importance, given their contemporary relevance. Approximately 2.78% of the respondents totally disagreed with the statement. Climate change has numerous effects on our lives. It also affects the way we live. Numerous studies have demonstrated that extreme climate phenomena possess a substantial potential to cause significant impacts on the energy system by disrupting electricity demand (Raju & Parveen, 2023). Approximately 36.79% of respondents totally agreed that carbon dioxide emissions are the leading cause of global climate change. The emission of greenhouse gases by human activities has caused global temperatures to rise, leading to climate change (Vijay et al., 2021). According to the study, the emission of greenhouse gases from industrial activities, fossil fuel combustion, and deforestation has contributed to a significant increase in global temperatures by 1 °C since 1900. Only a small fraction of 4.05% expressed their disagreement with the view that anthropogenic carbon emissions are the primary causes of global climate change.

3.3 Factors Affecting the Public on Energy-Efficient Labels

Implementing energy-efficient labeling has been found to positively impact people's lives, contributing to various benefits such as increased awareness of energy efficiency, cost savings, reduced environmental impact, and improved comfort and convenience. This study investigates the factors that influence the public's willingness to spend on energy-efficient labels to gain a deeper understanding of the variables that contribute to the effectiveness of energy labels in promoting energy

Table 6 A logistic regression on willingness to spend on energy-labeled appliance

Variables	Coefficient	Std. error	$P > \|z\|$
Age	0.0310	0.0107	0.004***
Years of education	−0.6948	0.3202	0.030**
Agree to purchase	−0.4044	0.2407	0.093*
Employment	1.3434	0.5588	0.016**
Environmental awareness	1.2387	0.2358	0.000***
Constant	3.8834	2.0766	0.061*
N	791		
Log-likelihood	−275.16918		
Pseudo R^2	0.0892		

*** significant at the 1% level; ** significant at the 5% level; * significant at the 10% level

conservation among consumers. The study employed a logit model to investigate the factors influencing the relationship between dependent and independent variables, where, in this case, the dependent variable is "willingness to spend" on energy-labeled appliances.

Table 6 indicates that age, employment status, and environmental awareness significantly impact the public's willingness to spend more money on energy-labeled appliances. The positive correlation observed between the variables indicates a significant increase in the probability of the public's willingness to spend on energy-labeled appliances. The age variable exhibits a statistically significant positive relationship at the 1% level, per findings. As an individual's age increases, the probability of willingness to spend more on energy-labeled appliances increases. According to Wang et al. (2020), the willingness to pay for energy-saving appliances can be influenced by many factors, including an individual's age, education level, income, and environmental consciousness, and among these factors, annual income has been identified as the most significant determinant. The variables "years of education" and "agree to purchase" indicate that the probability of buying energy-labeled appliances decreases as the covariate increases. In this situation, the coefficient of years of education (β, −0.6948), then $e^{-0.6948} = 0.49$, years of education is associated with a 51% (1–0.49 = 0.51), is indicative of a statistically significant decrease in the relative probability of willingness to spend more on energy-efficient appliances, by as much as 51%. This finding highlights the role of education in shaping consumer preferences for energy-efficient technology and the need for educational programs to promote energy conservation behavior.

3.4 Roles of Energy Efficiency (EE) Labeling for the Circular Economy

Energy efficiency labeling contributes significantly to the circular economy by promoting more sustainable consumption and production. The circular economy is an economic model that concentrates on reducing waste, reusing resources, and maximizing product lifespan. In a circular economy, considerations regarding energy selection and energy efficiency are of the utmost importance (Helena, 2022). Hence, EE labels contribute to these principles in multiple ways:

1. Providing Consumers' Awareness

Energy efficiency labels present consumers with information that is available to them regarding the amount of energy that is consumed by products such as home appliances and electronic equipment. Because of this awareness, consumers are able to make choices that are more informed, choosing products that are energy-efficient, typically have a longer lifespan, and have lower operating costs. The reduction in overall demand for energy and resources that results from consumers selecting these products is equivalent to providing indirect support for the circular economy.

2. Longer Product Lifespan

Products that are designed to conserve energy are typically created to higher quality standards, so they tend to have a longer lifespan than their less efficient counterparts. This helps to contribute to the circular economy by lowering the number of times that replacements are required and reducing the amount of waste produced. When customers are aware that a product is reliable and saves them money on their utility bills, they are less likely to get rid of it before its time.

3. Reducing Waste

Energy-efficient products typically need fewer repairs and replacements over their lifespan, which results in less waste in the form of rejected or outdated products. In addition, the lower energy consumption of such products lowers the burden that is placed on the environment by the processes of waste management and disposal.

4. Conserving Resources

The conservation of resources is closely connected to energy efficiency and other forms of resource efficiency. Products that require less energy to manufacture typically require less in the way of raw materials to do so. This is consistent with the goal of the circular economy, which is to reduce the consumption of resources and the environmental impact of manufacturing processes as much as possible.

5. Reducing Greenhouse Gas (GHG) Emissions

Products that are more energy-efficient use less energy, which in turn leads to lower levels of greenhouse gas emissions associated with the production of energy. This not only helps the environment, but it also helps the circular economy because it reduces the need for solutions that require a lot of resources to lessen the effects of climate change.

4 Conclusion

This study highlights the importance of energy labels in promoting sustainable development in Malaysia. Our empirical findings demonstrate that energy labels influence consumer choices, reduce energy consumption, and contribute to a more sustainable future. Energy labels can support the circular economy by offering information on the energy consumption and environmental impact of products, allowing stakeholders to make sustainable choices. Moving forward, policymakers, industry stakeholders, and the public must recognize the relevance of energy labels in policymaking and society. In the context of a circular economy, energy labels can serve as a catalyst for the adoption of products designed with durability, reparability, and recyclability in mind. Manufacturers, incentivized by consumer preferences reflected through energy labels, are encouraged to prioritize the use of recycled materials, design products for longevity, and facilitate ease of repair and disassembly. This shift toward energy-efficient and circular products not only reduces resource consumption and waste generation but also fosters innovation in product design and manufacturing processes, driving the circular economy agenda forward.

Moreover, energy labels can influence policy frameworks aimed at promoting circularity within industries. Governments can leverage energy labeling regulations to set minimum standards for energy efficiency and encourage the implementation of circular design principles among manufacturers. By incorporating criteria related to product lifespan, recyclability, and resource efficiency into energy label requirements, policymakers can incentivize businesses to embrace circularity as a core aspect of their operations. Additionally, energy labels can be integrated into broader policy initiatives such as extended producer responsibility (EPR) schemes, where manufacturers are held accountable for the entire lifecycle of their products, including end-of-life management. By linking energy labels to EPR regulations, policymakers can create a comprehensive framework that incentivizes manufacturers to design products with circularity in mind while also establishing mechanisms for effective waste management and recycling infrastructure. In this way, energy labels serve as a powerful tool for policymakers to drive systemic change toward a circular economy, fostering collaboration between governments, businesses, and consumers to create a more sustainable future.

Besides, by adopting the lessons from this study, we can take significant steps toward achieving our sustainability goals through effective energy labeling strategies and greater public awareness. Although this study has provided valuable insights into the relationship between energy labels and sustainable development, it also calls for further exploration. It is crucial to delve deeper into energy labeling through extensive research to advance toward a greener and more efficient future. The focus should be on analyzing the economic impact of energy labeling and evaluating its efficacy in various scenarios to refine strategies for sustainable development. Energy labeling provides a practical approach to work toward this objective, and it is the responsibility of everyone to pledge their commitment to sustainable development.

The study highlighted positive and significant factors that promote the public's willingness to spend more on energy labeling appliances, such as age, employment status, and environmental awareness. Effective awareness campaigns promoting energy labels are imperative for fostering public engagement in tackling environmental challenges. Energy labels are essential for advancing energy efficiency, mitigating greenhouse gas emissions, and curtailing environmental footprints. Maintaining a long-term commitment to raising awareness and promoting behavioral change is vital to guarantee lasting impact. To sustain continuous engagement with the public, sharing success stories, energy conservation tips, and updates on energy-efficient technologies and products is crucial. It is imperative to encourage the public to exchange their knowledge and experiences. Doing so can help establish a strong relationship with the public and form a community of like-minded individuals. Hence, it can foster a culture of learning and growth by facilitating an environment where individuals can share their expertise.

Energy labels are not just mere informative tags on appliances; they are catalysts for accomplishing the objectives of Sustainable Development Goal (SDG) 7: "Affordable and Clean Energy." By empowering the public with the knowledge to make energy-efficient choices, energy labels pave the way for energy conservation, reduced carbon emissions, and better access to sustainable and reasonable energy sources. As we work toward a future where everyone has access to clean and affordable energy, it is vital to identify the crucial role that energy labels demonstrate in driving positive change. They not only align with the principles of SDG 7 but also serve as a tangible tool in the collective journey toward a more sustainable and equitable world. Through informed choices and sustainable practices, we can confidently light the path toward a brighter, cleaner, and more energy-efficient future for all.

References

Abhijit, B., & Barry, D. S. (2003). Eco-labeling for energy efficiency and sustainability: A meta-evaluation of US programs. *Energy Policy, 31*(2), 109–123. https://doi.org/10.1016/S0301-4215(02)00012-5

Anna, C. R., Marta, R., Michele, G., & Claudio, F. (2018). Energy label directive: Current limitations and guidelines for the improvement. *Procedia CIRP, 69*, 674–679. https://doi.org/10.1016/J.PROCIR.2017.11.136

Cristian, H., de Lucinda, C. R., & Cardoso, A. R. (2020). Consumer response to energy label policies: Evidence from the Brazilian energy label program. *Energy Policy, 138*, 111207. https://doi.org/10.1016/J.ENPOL.2019.111207

Energy Commission of Malaysia. (2023). *Guideline for Energy Efficiency Label*. Retrieved from: https://www.st.gov.my/en/web/consumer/details/7/2

Farshid, B., Vahab, M., & Mohsen, J. (2013). Developing energy performance label for office buildings in Iran. *Energy and Buildings, 61*, 116–124. https://doi.org/10.1016/J.ENBUILD.2013.02.022

Guo, S., Lu, C.-P., Li, H., & Zhu, N. (2021). Influence mechanism of energy efficiency label on consumers' purchasing behavior of energy-saving household appliances. *Frontiers in Psychology, 12*, 711854. https://doi.org/10.3389/FPSYG.2021.711854
Haq, G., & Weiss, M. (2016). CO2 labelling of passenger cars in Europe: Status, challenges, and future prospects. *Energy Policy, 95*, 324–335.
Helena, R. G. (2022). The circular economy and energy. *CSR, Sustainability, Ethics & Governance*, 133–149. https://doi.org/10.1007/978-3-030-94293-9_7
Liang, G. (2011). *Environment-friendly adhesive label.*
Liuyang, Z., Meiting, J., & Jinpeng, L. (2011). Improvement of China energy label system to promote sustainable energy consumption. *Energy Procedia, 5*, 2308–2315. https://doi.org/10.1016/J.EGYPRO.2011.03.397
Mahlia, T. M. I., Masjuki, H. H., Saidur, R., Choudhury, I. A., & NoorLeha, A. R. (2003). Projected electricity savings from implementing minimum energy efficiency standard for household refrigerators in Malaysia. *Energy, 28*(7), 751–754.
Mahlia, T. M. I., Rahman, S., Yanti, P., & Masjuki, H. H. (2011). Role of energy guide labels in consumers purchase decision for household electrical appliances. *Energy Education Science and Technology Part A-Energy Science and Research, 27*, 95–104.
Masjuki, H. H., Saidur, R., Choudhury, I. A., & Mahlia, T. M. I. (2000). Factors effecting energy consumption of household refrigerator-freezers. In *2000 TENCON proceedings. Intelligent systems and Technologies for the new Millennium (Cat. No. 00CH37119)* (Vol. 2, pp. 92–96). IEEE.
Michalis, S., Dimitris, D., Tourkolias, C., & Areti, K. (2021). Efficient energy labelling: The impact of information content and style on product choice. *Energy Efficiency, 14*(58). https://doi.org/10.1007/S12053-021-09950-3
Mieko, F., Kazuhisa, T., Yukihiko, F., Nobuyuki, U., & Ryo, T. (2020). An experimental study of the effect of energy label design on the correct evaluation of buildings' energy performance and promotion of energy saving. *International Real Estate Review, 23*(1), 733–762.
Miranda, G., Abdel-Raheem, M. A., Mark, H., & Shaun, M. (2022). *Economics of energy efficient electrical appliances used in homes*. https://doi.org/10.1061/9780784483954.068.
Morteza, Z. O., Mohammadi-Ivatloo, B., Mohammadi-Ivatloo, B., Ahmadian, M. A. A., & Piran, M. J. (2020). A novel economic structure to improve the energy label in smart residential buildings under energy efficiency programs. *Journal of Cleaner Production, 260*, 121059. https://doi.org/10.1016/J.JCLEPRO.2020.121059
Paul, V., Paul, V., Peeters, J. R., Dirk, C., Paolo, T., Fulvio, A., Fabrice, M., Wim, D., & Joost, D. (2017). Ease of disassembly of products to support circular economy strategies. *Resources, Conservation and Recycling, 135*, 323–334. https://doi.org/10.1016/J.RESCONREC.2017.06.022
Rada, E. C., Ragazzi, M., Vincenzo, T., Giacomo, C., Luca, A., & Lucian-Ionel, C. (2018). Circular economy and waste to energy. *AIP Conference Proceeding, 1968*(1), 030050. https://doi.org/10.1063/1.5039237
Raju, B., & Parveen, K. C. (2023). Energy and climate change issues around CSUDH. *The CSU Journal of Sustainability and Climate Change, 2*(1). https://doi.org/10.55671/2771-5582.1017
Renate, S. (2017). *Purchasing energy-efficient appliances—To incentivise or to regulate?* (Vol. 4, pp. 215–234). Springer. https://doi.org/10.1007/978-3-319-50932-7_9
Shigeharu, A., & Yasuhiro, S. (2008). Start of uniform energy-saving label: Points of energy-saving measures for home electric appliances, etc. *IEEE Transactions on Electrical and Electronic Engineering, 3*(1), 32–36. https://doi.org/10.1002/TEE.20230
Signe, W., Bernadette, S., & Michael, S. (2015). Desired and undesired effects of energy labels—An eye-tracking study. *PLoS One, 10*(7), 1–26. https://doi.org/10.1371/JOURNAL.PONE.0134132
Sjors Van, W. (2019). The information value of energy labels: A quasi-experimental approach. Erasmus School of Economics, *Wrasmus University Thesis Repository, 2019-03-19.*

Tharinya, S., Hsiau, S. S., Lin, S.-M., Wongsapai, W., & Wu, J. C. (2017). Household energy consumption behaviour for different demographic regions in Thailand from 2000 to 2010. *Sustainability, 9*(12), 2328. https://doi.org/10.3390/SU9122328

Vijay, K., Deeksha, R., & Verma, K. (2021). Global climate change: The loop between cause and impact. *Global Climate Change*, 187–211. https://doi.org/10.1016/B978-0-12-822928-6.00002-2

Wang, X. E., Li, W., Song, J. N., Duan, H. Y., Fang, K., & Diao, W. Y. (2020). Urban consumers' willingness to pay for higher-level energy-saving appliances: Focusing on a less developed region. *Resources, Conservation and Recycling, 157*, 104760.

Yang, H. (2020). Energy-saving and environment-friendly paper label production device.

Yau, C. F. (2020). Legislative framework. In S. Kimura, P. Han, & S. M. Leong (Eds.), *Energy efficiency and conservation master plan of Cambodia. ERIA research project report FY2020 no.07* (pp. 33–45). ERIA.

Zhang, Y., Xiao, C., & Zhou, G. (2020). Willingness to pay a price premium for energy-saving appliances: Role of perceived value and energy efficiency labeling. *Journal of Cleaner Production, 242*, 118555.

Impact of Climate Change on Circular Economy Development and Ensuring Human Rights in Vietnam

Dang Thi Ha and Tran Thi Dieu Ha

Abstract Climate change is said to be one of the biggest challenges humanities must face in the twenty-first century, with Vietnam being one of the most vulnerable countries in the world. Weather phenomena such as droughts, floods, landslides, and storms have complicated developments. The risk of forest fires in the North Central and Central Highlands ecological regions is increasing. The risk of landslides in mountainous areas is affecting the agricultural and industrial economy, as well as impacting human rights and social justice (Ministry of Finance, 2022). Many adverse health effects due to climate change include infectious diseases, food and waterborne diseases, respiratory disorders and allergies, malnutrition, communal violence, and mental health problems. These environmental and health consequences threaten the assurance of human rights standards in Vietnam, including the right to life, the right to access safe food and water, the right to medical care, and equality of dignity and rights. According to some studies, it is more difficult to ensure human rights standards in developing countries, and disadvantaged groups (including the poor; members of minority groups, women, children, and elderly people; people with chronic illnesses and disabilities; people living in areas with high rates of climate-related illness; workers exposed to extreme heat or increased weather variability) are most vulnerable to the impacts of climate change (Tuana, 2012). In addition, they also suffer serious impacts from greenhouse gas emissions when living in urban areas or near industrial zones (Rajamani, 2018). In recognition of the critical nature of climate change, Vietnam is implementing solutions to deal with this situation, including shifting the agricultural economy that dependent on resources and heavy industry to a circular economy, and reduction of greenhouse gas emissions are strategic goals. However, minimizing the impact of climate change, maintaining economic growth, and ensuring human rights standards have never been an easy problem at present.

D. T. Ha (✉) · T. T. D. Ha
University of Law, Hue University, Hue, Vietnam
e-mail: hadt@hul.edu.vn; dangthiha@hueuni.edu.vn; hattd@hul.edu.vn; ttdieuha@hueuni.edu.vn

A. Heshmati, P. K. Nam (eds.), *Circular Economy and its Implementations in Southeast Asia*, Frontiers in South and Southeast Asian Development Research,
https://doi.org/10.1007/978-981-96-0539-2_9

Keywords Climate change · Circular economy · Human rights standards · Disadvantaged groups · Vietnam

1 Introduction

Vietnam is a low-income country, but it has become a bright spot of the regional economy in recent years. Vietnam has become one of the six countries in the Asian growth pole with the fastest economic growth in the world by 2030 (China, Vietnam, and Indonesia among the fastest-growing countries for the coming decade in New Harvard Growth Lab projections) with GDP growth rates over the years of 5.49% in 2020, 2.58% in 2021, and 8.02% in 2022.[1] In particular, when Vietnam joins the World Trade Organization and the Trans-Pacific Partnership Agreement, economic growth will increase enormously as international trade and foreign direct investment increase. These help Vietnam strengthen its goal of becoming a country with a developed economy and high income by 2045. However, this goal will be challenging to achieve when Vietnam is one of the countries heavily affected by climate change. According to German watch's "Long-Term Climate Risk Index Report 2019," Vietnam is considered one of the most vulnerable countries in the world to the impacts of climate change due to factors such as geographical location, economic development model, and population density. The decline in agricultural productivity and food security is a concern for developed and developing countries due to the impact of climate change.

Agriculture is more affected by unpredictable changes in temperature and precipitation as well as increased levels of floods and droughts compared to other industries (Ahsan, 2020; Chandio et al., 2020). *The Food and Agriculture Organization of the United Nations* also warned all countries to prioritize solutions to protect the agricultural sector against climate change and extreme *weather* (Food and Agriculture Organization United Nations Industry (FAO), 2018). Initial calculations show that Vietnam will lose 10 billion USD in 2020, equivalent to 3.2% of GDP, due to the impact of climate change. Without appropriate adaptation and mitigation measures, it is estimated that climate change will cause Vietnam to lose about 12–14.5% of GDP per year by 2050 and could put one million people in a "state of extreme poverty" by 2030 (World Bank, 2010).

When Vietnam develops economic development policies, the urgent challenge is to minimize people's vulnerabilities and ensure that economic growth is balanced with ensuring human rights standards. Environmental pollution, natural disasters, and epidemics are becoming more serious and leaving long-term consequences for Vietnam in building a legal corridor to ensure that people can live in a healthy environment (Levy & Patz, 2015). Along with the economic decline, the challenges that the Vietnamese government must face include food crisis, inadequate housing infrastructure, gender inequality, violence, migration and unemployment, shortages

[1] Report of the Ministry of Finance.

in medical and healthcare networks, mortality rate, environmental and occupational health issues, and social injustice. As a country with a long history of responding to and minimizing the impact of natural disasters, in sustainable development policies for 2020–2030, Vietnam prioritizes economic development based on agricultural and forestry resources while integrating human rights protection in economic and environmental policies and increasing solutions to respond to climate change. The ability to respond to disasters and reduce risks will increase as national and individual wealth increases (Tol, 2009), and prioritizing economic development is the appropriate solution to climate change. However, reality shows that economic development solutions prioritize development over the need to protect the environment and human rights standards. This requires the government to make efforts to balance economic goals and human rights standards.

2 Research Methods

Qualitative research methods are used to seek comprehensive assessments and analyze the impacts of climate change on the economy and its effects on public health. These assessments are the basis for considering the effectiveness and challenges of investing in developing a circular economy in Vietnam. At present, with the characteristics of a country with an average economy and dependence mainly on the agricultural economy, one of the economic factors susceptible to damage from natural disasters and climate change, what challenges will Vietnam face when building a circular economy while ensuring human rights?

Quantitative methods are used to collect information and data through numbers and statistics. This method aims to consolidate basic and general information in environmental protection policies in relation to economic development policies, circular economic development policy in Vietnam in the face of climate change, and balancing human rights factors. The results only allow qualitative analysis of the results.

3 Research Results

The Vietnamese government faces many threats and challenges in making reasonable decisions for economic development and environmental protection: the risk of resource depletion, environmental pollution, and climate change in the twenty-first century. In its development plans, the government always expresses its ambition for GDP growth while focusing on building a sustainable environment, and the application of circular economy development policies is a testament to that. Many studies have shown that the circular economy is a completely suitable model for Vietnam's current development goals (Chaudhry & Ruysschaert, 2007; Chau et al., 2023; Hai, 2019) instead of focusing on the traditional production model that is only interested in exploiting resources, produced and then thrown away after consumption, causing a large amount of waste to the environment. The circular economy focuses on managing and regenerating resources in a closed model to reduce waste by investing in new technology production lines. The circular economy is also the optimal solution for ensuring people's right to live in a healthy environment—one of the conditions to ensure human rights.

The key issue is the cognitive transformation toward a circular economy and its implementation in practice. It is important to consider whether there are enough grounds to determine whether the application of new technology in factory production lines is optimal in the current period and justify changes in economic development policies or not. In other words, does Vietnam have enough resources to solve the circular economy problems? What challenges is Vietnam facing when developing a national action plan on circular economy? Which option is most suitable to balance economic development and ensure human rights standards? The conclusion will address these questions and suggest management solutions in this area.

4 Impact of Climate Change on Economic Development and Human Rights Assurance

Climate change is considered one of humanity's biggest challenges in the twenty-first century due to its direct impact on the economy and other aspects of a country. Climate change is becoming an unpredictable and increasingly extreme phenomenon. Among Southeast Asian countries, Vietnam is one of the top 5 countries in the world most vulnerable and directly damaged by climate change. The Climate Change Vulnerability Index (CCVI) assesses the level of vulnerability due to the impacts of climate change in the next 30 years through 42 economic, social, and environmental factors in 193 countries—Vietnam is ranked 23rd out of 193 countries and is one of 30 countries at "very high risk" and directly vulnerable because of climate change (Nguyen The Chinh, 2016). According to statistics, in the period of 2011–2020, extreme climate caused serious economic damage, with total losses estimated at 229,958 billion VND (equivalent to 10 billion USD, according to 2022 exchange rates). Direct damage to public and private property: Vietnam loses an average of about 2.4 billion USD (equivalent to 0.8% of GDP) each year due to extreme weather problems. Furthermore, if calculated according to the cost of general environmental degradation, damage due to climate change is estimated at about 10 billion USD in 2020, equivalent to 3.2% of GDP (Ministry of Natural Resources and Environment, 2022).

The cause of this problem is considered to stem from the characteristics of Vietnam as a country with a long coastline and an economy dependent on agricultural and forestry production. According to Vietnam's "Socio-Economic Development Strategy" for the period 2021–2030, the country's economic transformation results will depend heavily on the effectiveness of natural resources management—

large reserves of agricultural resources, forests, and minerals have helped promote the development process in Vietnam (World Bank, 2010).

Regarding the agriculture and forestry sector: The weather in Vietnam in recent years has become increasingly unusual. Phenomena such as droughts, floods, landslides, and storms all have complicated developments and serious impacts, increasing the risk of forest fires in the North Central and Central Highlands ecological regions and the risk of landslides in mountainous areas (Cao Le Quyen, 2018). As a country whose economy depends mainly on agriculture, the impacts of climate change will affect more than 12% of Vietnam's agricultural land area and about 25% of its population. As an essential contributor to world food security—as a major exporter of rice, seafood, and coffee (UN Food Systems Summit, 2021)—climate change poses a severe threat to Vietnam's agricultural sector (World Bank Group and Asian Development Bank, 2020). According to estimates, agriculture accounts for about 20% of the country's GDP, and the negative impact of climate change and natural disasters on the agricultural sector could cause Vietnam's GDP to decrease by 0.7%–2.4% in 2050 (Stern et al., 2006).

For the fishing and seafood exploitation sector, when the coastal area's rainfall increases to 100 mm (0.1 m), the annual fishing output decreases on average from 0.98% to 2.2%. It may reduce 1.5% of next year's mining output (Ministry of Finance, 2022). A study result on the impact of climate change on the fisheries sector in ten Northern and North Central provinces shows the level of economic losses due to changes in temperature by 2050 (according to 2012 comparative prices, discounted 3%/year) will be 410 billion VND for the fisheries exploitation sector and 445 billion VND for the aquaculture sector, respectively (Ministry of Finance, 2022). Besides, it is predicted that Vietnam's salt industry will be the economic sector affected earliest, most directly, and most severely by sea level rise. Tens of thousands of hectares of salt production area will be at risk of gradually being wiped out over time (Dung & Sharma, 2017). The impact of climate change also has consequences for the processing and manufacturing industry due to the lack of raw materials for industry, especially raw materials for the food processing, textile, and garment industries because they cannot process supply from raw material areas in the Mekong Delta provinces—which are expected to suffer the most of severe flooding and climate change.

Climate change is causing widespread pollution in Vietnam, including pollution from exhaust gases and solid waste discharged from urban areas and heavy industry. Regarding wastewater pollution, statistics from the Ministry of Construction in November 2017 show that every day, 787 urban areas across the country with more than 33 million people discharge 3.1 million m3 of untreated wastewater (Clapp et al., 2018). In addition, 283 industrial parks, 615 industrial clusters, and nearly 3000 craft villages across the country discharge 0.7 million m^3 of wastewater every day, alarming when only more than 15% can be treated: waste pollution. Climate change has multifaceted negative impacts on enterprises' production and business activities. These include interruptions in production and business, reduced labor productivity, reduced revenue, disruption of transportation channels, increased production costs, stagnation of distribution networks, reduced product quality, damage to facilities material, manpower shortage, and lack of raw material supply. The economic recession is putting businesses in difficult times. Meanwhile, building a circular economy requires businesses to have abundant financial resources to apply high technology to production and waste treatment.

Along with economic impacts, climate change is increasing pressure on life and ensuring Vietnam's human rights standards. First, it is the right to live in a clean and healthy environment. Rising air temperatures, frequent heat waves, prolonged floods and rains, as well as extreme weather events tend to increase, increasing negative impacts on the living environment and human health. According to statistics, when the average temperature increases by 1.0 °C, the hospitalization rate for children aged 0–2 years increases by 3.4% and for children aged 3–5 years increases by 4.6% UNICEF (2021). Climate change indirectly impacts human health through sources of viral infectious diseases related to climate characteristics such as temperature, humidity, rainfall, and wind intensity. People in developing countries, such as Vietnam, are more vulnerable to natural disasters and climate change due to their dependence on agriculture, lack of resources, poor infrastructure, and unstable institutions. It is also the cause of poverty and disease (Bangalore et al., 2019).

Climate change increases the possibility of outbreaks and spread of diseases such as influenza A (H1N1, H5N1), influenza A (H7N9), diarrhea, cholera, dysentery, malaria, dengue fever, yellow fever, typhoid, Japanese encephalitis, severe acute respiratory syndrome (SARS), and plague (AFD, 2023). Some other groups of diseases such as high blood pressure, sunstroke/heatstroke, mental illness, chronic obstructive pneumonia, bronchial asthma, respiratory diseases, strokes, and complications due to heat waves or severe colds have also increased in the future.

Increasing gender inequality is one of the most significant impacts of climate change in Vietnam. Both men and women are adversely affected by climate change. However, in terms of gender-based coping capacity, women are a group that tends to be more vulnerable than men. Women account for 50.7% of Vietnam's total population, of which the proportion of women in urban and rural areas is 34.5% and 65.5%, respectively. Female workers in the agricultural sector account for 63%, more than men (57% as shown in Spencer et al. (2022); Binns et al., 2021). In rural areas, most women are involved in agricultural production activities including farming, animal husbandry, and postharvest work. They are the main ones responsible for unpaid care and housework in the family, so when climate change and natural disasters occur, it puts an even greater burden on women. It can cause harmful and long-term impacts on women's health (Binh, 2021). Climate change and natural disasters also increase the rate of death, injury, and disappearance of women. Therefore, the possibility of female workers being affected by climate change is also higher when the agricultural sector is one of the most adversely affected by natural disasters. For example, it has been repeatedly shown that women are more vulnerable to the adverse impacts of climate change than men, due to fundamental inequalities in access to land and assets security, lower levels of education, income, and limited participation in community decision-making processes, especially in low-income countries (UNDP, 2022; Nguyen Song Tung, 2015).

Human rights play an important role in international law and politics on climate change today. Legal scholars and political and moral philosophers are increasingly interested in human rights approaches to climate change. Studies on the impacts of climate change have begun to explore the links between human rights and climate change. Most studies believe climate change will be the root of human rights protection and solutions to climate change all focus on human rights issues. However, it will be difficult for Vietnam to develop solutions to promote the economy and ensure optimal human rights, especially in the context that the country has no financial reserves. Over 40 million people escaped poverty in 1990–2014, pushing the poverty rate from 50% in 1993 to 3% today (Vuong, 2014). Economic restructuring and reform have changed the proportion of sectors: the agricultural sector decreased from 40% to less than 20% of GDP, and industry and services increasingly accounted for a large proportion of the economy. Besides, Vietnam has completed many Millennium Development Goals and is on track to implement sustainable development goals successfully. Shifting an agricultural economy to an economy focusing more on industry and investment cooperation has brought the Vietnamese economy out of poverty and made it an average country worldwide.

However, the Vietnamese government still faces many challenges when pollution and industrial waste threaten human health and the environment. Vietnam has narrowed the area of agricultural land for cultivation to build factories and industrial parks. As a result, people are gradually losing their means of production, mainly agricultural land, and suffering the harmful effects of waste and industrial gas. According to statistics from the Ministry of Construction, each year, the total amount of domestic solid waste in Vietnam is nearly 16 million tons. The amount of plastic waste generated daily alone is estimated at nearly 18,000 tons. It is among the top 5 countries dumping trash into the sea, with a volume of about 1.8 million tons annually. Water pollution in urban areas: Every flood, waste, and garbage will attack residential areas, causing serious pollution (Chu, 2018). In the summer when the weather is hot, the air pollution index in large cities such as Hanoi and Ho Chi Minh City is always at alarming levels. These environmental problems are becoming even more severe due to the impacts of climate change and rapid population growth, especially in urban areas near coastal systems.

The world is facing a global economic recession. At the same time, climate change is increasingly extreme and unpredictable. Hence, a balanced solution between political stability, economic development, and ensuring human rights is necessary today for every country, including Vietnam.

5 Vietnam's Solutions in Dealing with Climate Change, Developing a Circular Economy and Ensuring Human Rights Standards

5.1 Solutions to Deal with Climate Change

Among solutions to maintain economic development, minimize the impact of climate change, and ensure human rights standards, building plans to respond to climate change is given top priority. Basic human rights will not be met if a country falls into poverty, and economic development will face many barriers if that country is not proactive in dealing with the complex developments of natural disasters.

Through different periods, Vietnam has always had a plan to respond to climate change corresponding to each period. Vietnam acknowledged the threat posed by climate change by ratifying the UNFCCC in 1994 and the Kyoto Protocol in 2002. This was also the period when the country's economy transformed from a command and centrally planned economy to a market economy. The period 2000–2010 was a difficult period for the economy. The Initial National Communication (INC) to the UNFCCC only explored the impacts of climate change and the necessary adaptation measures preliminarily and qualitatively (Chaudhry & Ruysschaert, 2008). A series of sector assessments have been carried out, and adaptation options have been identified. However, these sector assessments do not include socioeconomic analysis, and do not meet human rights standards (Chaudhry & Ruysschaert, 2008; Shaw, 2006). Plans for economic development and investment attraction were promoted in the following years, along with strategies to respond to climate change being deployed on a large scale. However, before 2020, Vietnam's policies were said to prioritize economic development above the need for environmental protection, so many localities had abolished or revised environmental planning and lowered environmental standards and requirements of environmental protection laws to attract investors (Dang Cong Cuong, 2020). In addition to being influenced by the mindset of exchanging environmental pollution for economic development, human rights issues have not been paid attention to when important documents such as the "Environmental Protection Law" have not been officially recognized as "human rights to the environment." Thus, in the policy to respond to climate change in these stages, Vietnam's economy mainly focuses on developing based on available resources and the salvaged industrial model without focusing on separating economic activity from consuming finite resources and limiting waste generation while transitioning to renewable energy sources.

Vietnam officially recognized human rights to the environment (in other words, the right to live in a clean environment) through a separate regulation when passing the Law on Environmental Protection in 2020. Policy to deal with Climate change focuses on three main directions:

First, in the strategy for sustainable development and response to climate change in the current period, Vietnam builds a strategy for developing a circular economy and responding to climate change in a human-centered way. The plan focuses on creating a greener environment by reducing pollution, recycling waste, and stopping the use of coal power by 2040; Preserve natural resources and natural resources, restore primary forests; End deforestation by 2030 and contribute to a 30% reduction in global methane emissions compared to 2020 levels.

Second, Vietnam focus on building infrastructure to respond to climate change in areas at high risk of being affected by climate change; Implement policies to ensure food security, and reduce waste and industrial emissions through modern treatment technology systems; Industrial development in the direction of developing a green industrial enterprise model.

Third, increase the community's ability to proactively respond through building a natural disaster forecasting system: According to a report from the General Department of Meteorology and TV, by April 2020, the hydrometeorological station network had 1640 stations/points measurement, including 202 meteorological stations; 14 automatic radiation measuring stations; 29 agricultural meteorological stations; 01 global climate monitoring station in Pha Din; 782 independent automatic rain measuring points; 404 hydrological stations (including 162 stations that are not part of the Hydrometeorological Station Network Development Plan invested in serving automatic hydrological monitoring of inland areas for the Mekong Delta); 27 oceanographic stations; 179 air and water environmental monitoring stations/points (including 91 salinity measurement points); 27 overhead meteorological and weather radar stations along with 18 lightning location stations (General Department of Hydrometeorology, 2020).

The sustainable development plan for the period 2020–2030 focuses resources on green transformation, building a national action plan on circular economy with the goal of minimizing the impact of climate change to ensure human rights. In particular, to limit risks and challenges for women in the reform period, Vietnam needs to proactively legalize international treaties and laws on gender equality in response to climate change. Women's rights need to be integrated into policies on natural disaster prevention and control, sustainable poverty reduction, green growth, circular economy, and vocational education development (Nguyen Ha, 2021). In addition, Vietnam prioritizes gender equality and empowering women and girls in climate change (Ngan Anh, 2022; Razafindrabe, 2014). Policies to deal with climate change are child-centered with three orientations: develop policies to respond to climate change that are child-centered; raise awareness about the impact of climate change on children and strengthen international cooperation; region in responding to the impacts of climate change on children (Huang et al., 2022). This demonstrates the government's efforts to meet human rights standards integrated with economic development in the global economic recession.

5.2 Evaluation

It can be said that Vietnam's current policies have taken timely steps to adapt to climate change.

Developing policies and laws to ensure human rights standards is extremely necessary and meaningful. These contents have been presented in the above section. In this section, we want to evaluate the effectiveness of solutions based on climate change response policies and Vietnam's current economic development policies.

First. Governments need to ensure that human rights are considered in developing and implementing climate change mitigation and adaptation measures. The green growth strategy recently approved by the Prime Minister demonstrates the determination of the Vietnamese government for growth with clean technology, sustainable development, and protection of the environment and natural resources. However, most businesses encounter difficulties in green transformation because of institutional factors. According to estimates by the World Bank (WB), to complete the goal of achieving net zero emissions by 2050 within the framework of Vietnam's commitment at the 26th Conference of the Parties to the United Nations Framework Convention on Climate Change (COP26), Vietnam will need up to 380 billion USD—this requires a huge resource. Calculating capital needs specifically in the energy sector, the funding needed in the energy transition process according to the Power Plan for the period 2021–2030 amounts to nearly 135 billion USD, equivalent to 13.5 billion USD/year (Government, 2023). Therefore, it will be very difficult if Vietnam only relies on corporate capital. Furthermore, as observed over the past 10 years, capital accumulation and internal financial strength of businesses are still quite thin. Besides, many banks consider green projects as normal projects, making it difficult to encourage investors because interest rates are not preferential, it is difficult to compete, and the processing time is also longer.

Vietnam aims to develop a circular economy in the period 2021–2030 with an average gross domestic product (GDP) growth rate of about 7%/year; GDP per capita at current prices by 2030 will reach about 7500 USD, an increase compared to the average in the 2016–2020 period of 1.1% (the average in the 2016–2020 period is about 5.9%/year, according to Government (2023)). To achieve the above goals, Vietnam prioritizes shifting the agricultural economy to industry and investing in science and technology in agriculture and industry. Advanced and modern production technology will free up labor in harsh weather situations, minimizing emissions and waste into the environment. However, the economic development strategy during this period still depended on agricultural resources, forestry, and mineral exploitation. Economic shifts toward applying science and technology and innovation are still low, with a lack of highly skilled human resources capable of using technology for agriculture and industry (Tien et al., 2021). Compared to other countries, financial resources from the state budget to invest in science and technology in Vietnam are still limited: spending nearly 0.6% of GDP in 2020, 1.01% of GDP in 2022, and 0.82% of GDP in 2023, according to the statement. According to the Minister of Finance, spending on science and technology has gradually decreased since 2017, and this is an alarming number. Meanwhile, the R&D spending rate of the European Union (EU) in 2017 reached 2.03% of GDP; China reached 2.19% of GDP; Japan reached 3.26% of GDP; Singapore reached 1.95% of GDP; and Malaysia reached 1.95% of GDP (Ministry of Finance, 2022). For

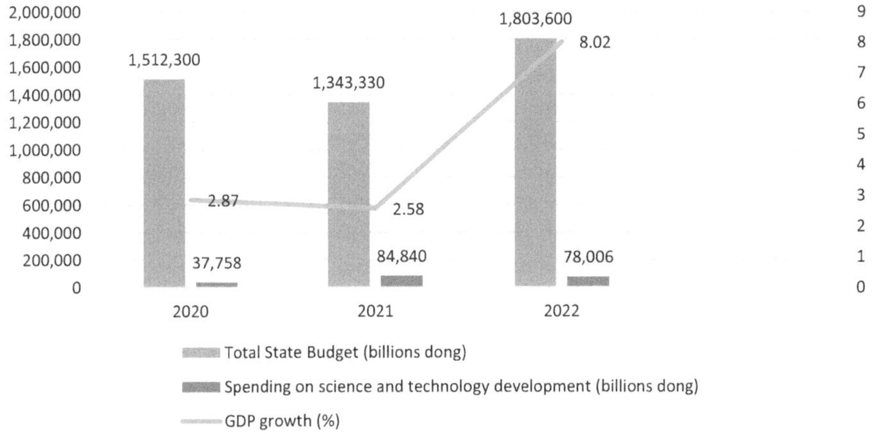

Fig. 1 Correlation chart between total state budget, spending on science and technology development, and GDP growth

businesses, green development and circular economy are gradually becoming a global trend as a positive solution to reduce greenhouse emissions and improve the resilience and creativity of the economy, toward the goal of carbon neutrality and sustainable development. This is shown in Fig. 1:

It can be seen that Vietnam's GDP growth rate in the period 2020–2021 is not high, fluctuating at 2.87% and 2.58%, but will witness a jump in 2022 when it reaches 8.02%. However, this number is only a "step forward" compared to the growth rate in 2021, thanks to the Vietnamese government focusing all resources on reviving the economy after the COVID-19 pandemic. With the total state budget and growth rate not too high in recent years, it can be said that Vietnam does not have enough resources to invest maximally in science and technology—one of the decisive factors for a circular economy. This is demonstrated in the data on investment levels in science, technology, and technical innovation in 2020, 2021, and 2022, which always remain at a low level. At any stage, spending on science and technology development only accounts for 0.02% (2020) to 0.04% (2022) of the total state budget. This data shows that although a higher GDP growth rate often indicates a stronger economy, it does not guarantee higher budget allocation for science and technology development—because capital allocation for science and technology, in this case, is affected by the government's priorities when focusing resources on post-pandemic medical care, infrastructure construction, and budget constraints.

Vietnam currently does not have a legal framework or overall policy on green credit activities. At the same time, there is a lack of tools and monetary policies to encourage banks to participate in financing. The regulations on green bond issuance also do not have specific instructions in cases where banks are the issuing organizations to mobilize capital. Vietnam does not have organizations that provide consulting services, appraisal, and assessment of social environmental risks and project green criteria. Regarding the treatment of solid waste generated from daily activities in urban areas, Vietnam does not have a mechanism to provide incentives, support for investment capital, or support for the consumption of processed products for waste treatment projects. The number of domestic solid waste treatment projects that receive loans from preferential capital sources is very small.

Specific mechanisms on incentives for recycling, reusing waste, recovering energy from household solid waste treatment, and calling for investment in solid waste treatment are lacking and inconsistent. Reusing waste resources is a strange concept in Vietnam today. This development trend has been forming and creating new game rules for trade and investment globally. However, it must also be recognized that the transformation of export manufacturing enterprises toward green and sustainable goals still faces many practical difficulties. Converting to green production and green export requires large investment capital and high technology that not all businesses have enough capacity to pursue and meet. In addition, new investment businesses following the circular economy model also need good, qualified experts to help the business orient its development. However, currently, the issue of training in the field of circular economy model design has not been focused on and promoted effectively.

This means it will be very difficult to reduce the proportion of heavy industry in the near future, and the goal of reducing greenhouse gas emissions will be difficult to achieve.

Second. Vietnam has not invested properly in minimizing the harmful effects of climate change. With many proposed solutions, Vietnam does not have a separate budget for dealing with climate change, with its allocation mixed with the environmental budget. Since 2005, the environmental budget (including climate change) has been regulated at no less than 1% of total budget expenditure. Determining the total costs for implementing climate change mitigation and adaptation is still relatively limited. The main statistical data and reports are total cost data for responding to climate change in general. Meanwhile, there is a difference between the collection of environmental protection taxes and the level of spending on environmental protection.

Since 2015, the environmental protection tax has contributed significantly to revenue. It is estimated that the expected revenue from environmental protection tax in 2019 is 68,926 billion VND, meaning that in just 7 years, revenue has increased more than six times, equivalent to an average annual revenue growth of 35% (Truong, 2021). While revenue from this tax source increases rapidly, spending on environmental protection is carried out in a trickle. Specifically, in 2012, revenue from environmental protection tax was 11,160 billion VND, which increased to 42,393 billion VND in 2016, equivalent to an increase of four times within 5 years. On the contrary, the amount spent on environmental protection increased insignificantly, from 9000 billion VND in 2012 to 12,290 billion VND in 2016. Thus, while environmental tax revenue increased 4 times, spending on environmental protection increased less than 1.4 times (*Summary report and assessment of implementing the Environmental Protection Tax Law in 2012–2016*). The level of state budget

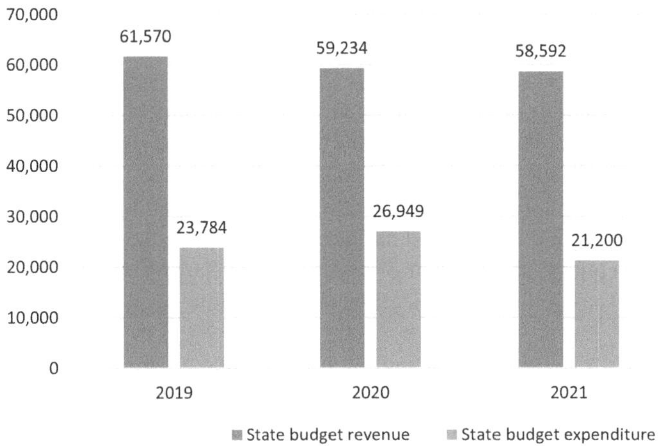

Fig. 2 Correlation chart of state budget revenue and expenditure for environmental protection (Unit: billion VND)

revenue and expenditure for environmental protection also has a significant "difference" in 2020, 2021, and 2022 as in Fig. 2:

In 2019, the budget revenue was 61,570 billion VND, 2.6 times higher than the expenditure of 23,784 billion VND. The year 2020 witnessed a budget revenue for environmental protection of 59,234 billion VND, 2.2 times higher than the expenditure of 26,949 billion VND. This difference continues to increase sharply in 2021 when it reaches a difference of 2.8 times with a revenue of 58,592 billion VND and an expenditure of 21,200 billion VND. In general, the level of budget revenue is always more than double the level of budget expenditure for environmental protection activities, showing an extremely large mismatch. This reality requires positive investment changes for general environmental protection and minimizing the harmful effects of climate change in particular.

Third. Urban waste is a terrible threat to the living environment—one of the important factors in evaluating the effectiveness of building a circular economy. Thousands of tons of solid waste are generated daily in industrial production and human life due to rapid urbanization, migration, and industrialization not only in Vietnam but in all developing countries. This situation is causing many serious problems for the government in the process of environmental management. In the current context of building a circular economy in Vietnam, many approaches, solutions, and programs have been researched, proposed, and implemented over the past decades. However, most have stopped at the pilot program and are generally not capable of being replicated, so it has not been widely applied.

The most challenging problem of the domestic solid waste management system is that waste classification at source has not been applied or fully implemented (Hoa & Matsuoka, 2017). The reason is not only due to limitations in collection equipment, human resources, and working efficiency but also due to lack of investment in infrastructure and limited awareness of the people (Van Tung et al., 2021). Construction waste and bulky waste being gathered and dumped are widespread, while the management and treatment mechanism is not drastic. Solid waste is an existential threat to the environment and the community's right to live in a healthy environment. Besides focusing on treatment plans, having a solid waste management mechanism must be carried out drastically and synchronously, which Vietnam needs to do better (Pushkareva, 2020). Solutions to raise public awareness have not been focused, and the management mechanism for this issue still has many gaps, putting the problem of solid waste treatment in context: the amount of waste is increasing, while technology, infrastructure, and scale of waste collection and treatment do not meet the needs.

Fourth. In its current strategies to deal with climate change, Vietnam places a strong emphasis on prioritizing children at the center of its response efforts. However, solutions to ensure children's rights against the impacts of climate change only stop at raising children's awareness of the impacts of climate change at school and some medical solutions. There are few solutions to cope with other impacts of climate change on children such as plans to prevent diseases for children before and after natural disasters, guaranteed vaccination regime, malnutrition in children, consequences of immigration, and domestic violence (Miller & Dun, 2019; Resurrections, 2012). Climate change also exacerbates gender imbalances in health, land ownership, and women's lack of power in families and communities (Phan et al., 2019). Although the Law on Gender Equality in 2006 and the Law on Prevention and Control in 2022 have created a legal corridor to protect women's rights and reduce domestic violence and gender imbalance, the issue of ensuring the right to life is for women has not yet improved. According to data from the General Statistics Office, in Vietnam, women often work in low-income occupations or vulnerable jobs, with little access to production, education, skills, and job opportunities than men. Vietnam's gender income gap also tends to widen. In 2021, the average monthly income of male workers is 7.7 million VND, while the average monthly income of female workers is 5.7 million VND (General Statistics Office, 2022). This shows that efforts to improve gender inequality in Vietnam have not been effective and have not met the most basic rights of women in the face of climate change.

6 Conclusions and Suggestions

As analyzed above, solutions to cope with climate change in Vietnam will be very difficult to implement when Vietnam does not have enough internal resources to balance policies addressing both climate change and economic development. Economic development is prioritized with human rights as an integral part of the overall strategy. The 2015 Lancet Commission on Health and Climate Change identified the policy needed to deal with the impacts of climate change to "ensure the achievement of the highest achievable health standards for people around the world" (Levy & Patz, 2015). Because climate change causes negative effects on public health, economic development needs to consider the right to live in a healthy environment as a basic right that needs to be protected as a priority. Among the many solutions, within the scope of this article, we offer solutions to minimize the impact of climate change. These are internal solutions, based on existing conditions, and can be implemented at present.

From the state, the government's policy actions must be compatible with the set goals. NGOs and humanitarian organizations need to hold governments accountable for protecting and promoting these human rights. The government has to evaluate and reconsider the serious nature of climate change concerning human rights issues and needs to have a separate and long-term response plan to climate change and reasonable financial resources to minimize the harmful effects of climate change and protect the environment. When human rights violations occur, governments should develop and deploy monitoring systems to detect and respond to any further violations. Governments should coordinate multi-sectoral engagement across agencies and organizations, ensuring a focus on protecting vulnerable populations.

To build a healthy living environment and ensure people's rights to life and health, treating industrial waste and solid waste in urban areas is one of the important issues for low-income countries. Vietnam is not a country with abundant financial reserves. However, the issue of developing a green economy requires a country with strong enough resources to invest in high technology in agricultural production and production light industry, waste treatment. Businesses require substantial funding to build modern production systems and invest in advanced waste and emission treatment processes. Therefore, the issue of industrial waste treatment in Vietnam at present is not a problem that can be resolved quickly.

The goal of building a green economy will be difficult to achieve if Vietnam does not promote its internal strengths and use community power in environmental protection. Firstly, Vietnam needs to reduce the amount of household solid waste. To do that, the government needs to clearly explain how to classify waste and quickly deploy a policy requiring people to segregate their waste according to organic waste, recyclable waste, reusable waste, food waste, and other types of solid waste. Although the government has written instructions on this issue, the instructions are not clear, and most people cannot distinguish between recyclable waste, reusable waste, and other types of waste, so the process of classifying waste is difficult (Hoang & Fogarassy, 2020). Secondly, in the context of mechanisms, policies, and models that have not attracted private investors to invest in waste treatment, it is necessary to deploy households to classify waste at source uniformly and quickly. The two issues mentioned above need to be implemented quickly and thoroughly.

The state should enhance cooperation with nongovernmental organizations and medical organizations. A recent study shows that the high cost of the current healthcare system in Vietnam leads to fewer people seeking medical care, creating conditions for chronic diseases to develop. If the government does not manage and handle this issue well, 40% of Vietnamese people must pay for medical treatment out of their own pockets (Hung et al., 2022; Roth, 2017). Social security in Vietnam currently does not fully cover the medical examination and treatment costs for all patients. Developing a community healthcare program that integrates nutritious meal planning with medical care is crucial. However, in the current climate change response strategies, the importance of public healthcare, particularly in the area of nutrition-focused meal design, has not yet been addressed. Many studies show that designing nutritious meals is the best way to improve health and prevent disease (McClements, 2020). Climate change is causing consequences for public health—a research campaign on designing meals suitable for Vietnamese people's physical health is a useful measure in policies to deal with climate change and ensure the protection of current human rights values. NGOs work in different fields to achieve cultural, social, philanthropic, professional, and industrial goals. With a high sense of responsibility toward community issues, they strive to support problem-solving and promote society. Considered the social capital of governments, NGOs can take advantage of maximizing their capacity to collaborate with the public sector and provide different social groups with necessary services in situations where the government cannot meet the needs of all social groups due to a lack of technical or administrative capacity or, more importantly, lack of financial resources (Rajabi et al., 2021). Collaboration with nongovernmental organizations in community healthcare and building appropriate nutritional diets are a necessary choice.

Climate change is a challenge for every country globally, threatening most people and human rights in the world. The adverse consequences of climate change will become increasingly more severe. Tackling climate change is a top health and human rights priority, and action cannot be delayed. Mitigation and adaptation measures must protect and promote human rights.

References

Ahsan, M. M. (2020). Strategic decisions on urban built environment to pandemics in Turkey: Lessons from COVID-19. *Journal of Urban Management, 9*(3), 281–285.

Bangalore, M., Smith, A., & Veldkamp, T. (2019). Exposure to floods, climate change, and poverty in Vietnam. *Economics of Disasters and Climate Change, 3*, 79–99.

Binh, V. N. (2021). Gender equality and women's empowerment in climate change adaptation and disaster risk reduction–laws, policies and practices in Vietnam.

Binns, C. W., Lee, M. K., Maycock, B., Torheim, L. E., Nanishi, K., & Duong, D. T. T. (2021). Climate change, food supply, and dietary guidelines. *Annual Review of Public Health, 42*, 233–255.

Cao Le Quyen (2018). Impact of climate change on the forestry sector and proposed adaptation measures, propaganda magazine.

Chandio, A. A., Jiang, Y., Rehman, A., & Rauf, A. (2020). Short and long-run impacts of climate change on agriculture: an empirical evidence from China. *International Journal of Climate Change Strategies and Management, 12*(2), 201–221.

Chau, E. C. T., Kwong, T. C., Pang, C. K., Chan, L. T., Chan, A. M. L., Yao, X., ... & Kwan, Y. W. (2023). A novel probiotic-based oral vaccine against SARS-CoV-2 omicron variant B. 1.1. 529. *International Journal of Molecular Sciences, 24*(18), 13931.

Chaudhry, P., & Ruysschaert, G. (2007). *Climate change and human development in Viet Nam: A case study for the human development report 2007/2008*. UNDP.

Chaudhry, P., & Ruysschaert, G. (2008). *Climate Change and Human Development in Viet Nam: A case study for how change happens.*

Chu, T. T. H. (2018). Environmental pollution in Vietnam: Challenges in management and protection. *Journal of Vietnamese Environment, 9*(1), 1–3.

Clapp, J., Newell, P., & Brent, Z. W. (2018). The global political economy of climate change, agriculture and food systems. *The Journal of Peasant Studies, 45*(1), 80–88.

Dang Cong Cuong (2020). Pháp luật đảm bảo quyền được sống trong môi trường trong lành ở Việt Nam, I*ndustry and Trade Magazine*, vol5.

Department of Climate Change, Ministry of Natural Resources and Environment in collaboration with the French Development Agency (AFD) (2023). *Report: "National Climate Change Impacts and Adaptation-Final Report.*

Dung, P. T., & Sharma, S. (2017). Responding to climate change in the agriculture and rural development sector in Vietnam. In *Redefining diversity & dynamics of natural resources Management in Asia*, (Vol. 2, pp. 13–25). Elsevier.

Food and Agriculture Organization United Nations Industry (FAO), 2018

General Department of Hydrometeorology. (2020). *Unify statistical data and update information of hydrometeorological stations 8Official dispatch no. 562/TCKTTV-QLML dated April 27, 2020 of the general director of the general Department of Meteorology and TV*. General Department of Hydrometeorology.

General Statistics Office. (2022). *Report on the recovery of the labor market employment after the Covid 19 pandemic, the third quarter of 2022.*

Government. (2023). *Resolution No. 90/NQ-CP dated June 16, 2023.*

Hai, D. M. (2019). Current situation and solutions for the environmental pollution in Vietnam. *Journal of Mechanical Engineering Research and Developments, 42*, 06–10.

Hoa, N. T., & Matsuoka, Y. (2017). The analysis of greenhouse gas emissions/reductions in waste sector in Vietnam. *Mitigation and Adaptation Strategies for Global Change, 22*, 427–446.

Hoang, N. H., & Fogarassy, C. (2020). Sustainability evaluation of municipal solid waste management system for Hanoi (Vietnam)—Why to choose the 'Waste-to-Energy'concept. *Sustainability, 12*(3), 1085.

Huang, H., Huyskens, M. H., Yin, Q. Z., Cawood, P. A., Hou, M., Yang, J., et al. (2022). Eruptive tempo of Emeishan large igneous province, southwestern China and northern Vietnam: Relations to biotic crises and paleoclimate changes around the Guadalupian-Lopingian boundary. *Geology, 50*(9), 1083–1087.

Hung, T. M., Van Hao, N., Yen, L. M., McBride, A., Dat, V. Q., van Doorn, H. R., & Schultz, M. (2022). Direct medical costs of tetanus, dengue, and sepsis patients in an intensive care unit in Vietnam. *Frontiers in Public Health, 10*, 1665.

Levy, B. S., & Patz, J. A. (2015). Climate change, human rights, and social justice. *Annals of Global Health, 81*(3), 310–322.

McClements, D. J. (2020). Future foods: A manifesto for research priorities in structural design of foods. *Food & Function, 11*(3), 1933–1945.

Mehmood, U., Imran, A., Abid, A., & Tariq, S. (2022). ul Haq. Environ. Res, 19

Miller, F., & Dun, O. (2019). Resettlement and the environment in Vietnam: Implications for climate change adaptation planning. *Asia Pacific Viewpoint, 60*(2), 132–147.

Ministry of Finance (2022). Assessing climate change in Vietnam.

Ministry of Finance (2022). Climate Change Assessment Report in Vietnam.

Ministry of Natural Resources & Environment. (2022). *Report: "The urgent situation of the Mekong Delta - Environmental and social adaptation strategy to 2050.*

Ngan Anh (2022). Việt Nam dành nhiều ưu tiên về bình đẳng giới và trao quyền cho phụ nữ, trẻ em gái trong biến đổi khí hậu, https://nhandan.vn/.
Nguyen Ha (2021). Việt Nam lồng ghép bình đẳng giới trong ứng phó với biến đổi khí hậu, https://laodong.vn, access September 28, 2021.
Nguyen The Chinh (2016). Proceedings of the International Scientific Conference "Kinh tế Việt Nam trong trung hạn: triển vọng và một số ảnh hưởng của yếu tố môi trường", Institute of Strategy and Policy on Natural Resources and Environment, Ministry of Natural Resources and Environment.
Nguyen Song Tung (2015). State-level science and technology project "Nghiên cứu xây dựng cơ chế, chính sách liên kết vùng trong việc ứng phó với biến đổi khí hậu ở Việt Nam, Code BCC-30 under "Science and technology program to serve the national target program to respond to climate change", Code: KHCN-BĐKH/11-15.
Phan, L. T., Jou, S. C., & Lin, J. H. (2019). Gender inequality and adaptive capacity: The role of social capital on the impacts of climate change in Vietnam. *Sustainability, 11*(5), 1257.
Pushkareva, L. (2020). *Implementation of the law on solid waste management in Vietnam today: Necessity, problem and solutions*. Available at SSRN 3663174.
Rajabi, M., Ebrahimi, P., & Aryankhesal, A. (2021). Collaboration between the government and nongovernmental organizations in providing health-care services: A systematic review of challenges. *Journal of Education and Health Promotion, 10.*
Rajamani, L. (2018). Human rights in the climate change regime. The human right to a healthy environment, 236–251.
Razafindrabe, B. H., Kada, R., Arima, M., & Inoue, S. (2014). Analyzing flood risk and related impacts to urban communities in Central Vietnam. *Mitigation and Adaptation Strategies for Global Change, 19*, 177–198.
Resurreccion, B. P. (2012). The gender and climate debate: More of the same or new pathways of thinking and doing? In *Human security and climate change in Southeast Asia* (pp. 95–111). Routledge.
Roth, L. (2017). *Leading causes of death in Vietnam*.
Shaw, R. (2006). Community-based climate change adaptation in Vietnam: Inter-linkages of environment, disaster, and human security. *Multiple Dimensions of Global Environmental Change*, 521–547
Stern, N. (2006). *Stern review: the economics of climate change*. HM treasury.
Spencer, S., Samateh, T., Wabnitz, K., Mayhew, S., Allen, H., & Bonell, A. (2022). The challenges of working in the heat whilst pregnant: Insights from Gambian women farmers in the face of climate change. *Frontiers in Public Health, 10*, 152.
Tien, N. H., Ngoc, N. M., & Anh, D. B. H. (2021). Current situation of high-quality human resources in FDI enterprises in Vietnam-solutions to attract and maintain.
Tol, R. S. J. (2009). The economic effects of climate change. *Journal of Economic Perspectives, 23*(2), 29–51.
Tuana, N. (2012). Climate change and human rights. In Handbook of human rights (pp. 410–418). *Routledge International Journal of Multidisciplinary Research and Growth Evaluation, 2*(1), 31–38.
Truong, D. D. (2021). Economic instrument in environmental management: A case study of environmental protection tax in Vietnam. *Environment and Ecology Research, 9* (5), 300–313.
UNDP. (2022). *Human rights in climate change*. File truy cập: file:///C:/Users/Admin/Downloads/VIE_Human%20Rights%20and%20Climate%20Change.pdf.
UN Food Systems Summit (2021). Dismantling democracy and resetting corporate control of food systems. *Frontiers in Sustainable Food Systems, 5*, 661552.
UNICEF (2021). Report of the United Nations Children's Fund on "Action against the climate crisis".
Van Tung, T., Thao, N. T. P., Hieu, T. T., Le Thanh, S., Braunegg, S., Braunegg, G., & Schnitzer, H. (2021). Waste treatment and soil cultivation in a zero-emission integrated system for catfish farming in Mekong delta, Vietnam. *Journal of Cleaner Production, 288*, 125553.

Vuong, Q. H. (2014). *Vietnam's political economy in transition (1986–2016)*. Stratfor-The Hub: International Perspectives.

Women, U. N. (2016). Global guidance on addressing school-related gender-based violence. UNESCO Publishing.

World Bank. (2010). *World development report 2010: Development and climate change*. World Bank.

World Bank Group and Asian Development Bank, 2020.

EU's Legal Framework for Secondary Raw Materials and Recommendations for Vietnam: Toward Sustainable Resource Management and Circular Economy

Huynh Phan Nhu Ngoc and Nguyen Ngoc Phuong Dinh

Abstract This study thoroughly explores the European Union's legal framework for secondary raw materials and offers customized recommendations to empower Vietnam in the pursuit of sustainable resource management and the adoption of circular economy principles. The paper delves into key EU directives, regulations, and policies, emphasizing their role in optimizing secondary raw materials. A comparative analysis of Vietnam's current regulations is provided, identifying opportunities for alignment. Additionally, the article showcases real-world examples such as IKEA's sustainable approach to recycled materials, a theoretical framework for circular economy implementation, and methodology for circular economy transition using assessment tools for secondary raw materials. Detailed statistical data on the circular material use rate in the EU and a case study on Vietnam's initial success with Ecoslag further enrich the content. This comprehensive analysis paves the way for Vietnam's transition toward a sustainable and circular economy.

Keywords Circular economy · Secondary raw materials · European Union (EU) · Sustainable resource management · Recycling rates · Waste generation · Environmental impact · Vietnam · Legal framework · Circular material use rate · Resource efficiency

1 Introduction

The notion of a circular economy has garnered substantial global attention as a compelling remedy for addressing urgent environmental and resource-related issues. With economies experiencing continuous growth, the conventional linear approach of "take, make, dispose" has resulted in excessive resource utilization, the generation of significant waste, and widespread ecological degradation. In light of these challenges, the European Union (EU) has positioned itself as a frontrunner on the world

H. P. N. Ngoc (✉) · N. N. P. Dinh
University of Economics, Ho Chi Minh City, Vietnam

A. Heshmati, P. K. Nam (eds.), *Circular Economy and its Implementations in Southeast Asia*, Frontiers in South and Southeast Asian Development Research,
https://doi.org/10.1007/978-981-96-0539-2_10

stage, advocating for a circular economy and sustainable resource management through a comprehensive legal framework.

1.1 Background and Importance of Secondary Raw Materials

The linear economy's shortcomings have spurred interest in transitioning toward a more sustainable and circular approach to resource management. In a circular economy, the value of products, materials, and resources is preserved for as long as possible, and waste is minimized through reuse, remanufacturing, recycling, and other innovative practices. A central element of circularity is the concept of secondary raw materials, which refers to materials recovered from waste and used as inputs for new products or processes.

Secondary raw materials play a pivotal role in reducing the extraction of virgin resources and lowering greenhouse gas emissions associated with traditional production processes. By promoting the use of secondary raw materials, the circular economy enables resource efficiency, reduces environmental impact, and enhances overall economic resilience. The EU recognizes the immense potential of secondary raw materials in mitigating environmental degradation and enhancing resource security, and as a result, has devised a comprehensive legal framework to foster their optimal utilization.[1]

1.2 The Circular Economy Concept and Its Relevance to Sustainable Resource Management

The circular economy framework aligns seamlessly with the broader concept of sustainable resource management. It embodies the principles of preserving resources, minimizing waste, and fostering sustainable production and consumption patterns. At its core, the circular economy envisions a shift away from the traditional "take, make, dispose" mindset and toward a regenerative approach that prioritizes resource efficiency and long-term viability.[2]

Through circular economy practices, materials are continually cycled back into the economy, reducing the reliance on finite resources and mitigating environmental impacts. This approach fosters resilience against resource scarcity, volatile commodity prices, and environmental risks. Furthermore, the circular economy provides a pathway to decoupling economic growth from resource consumption, offering a sustainable blueprint for future generations.

---

[1] Watkins & Meysner (2022).

[2] Watkins & Meysner (2022).

1.3 Rationale for Studying the EU's Legal Framework and Its Applicability to Vietnam

The EU has emerged as a trailblazer in implementing circular economy principles, with an extensive legal framework dedicated to advancing sustainable resource management. By examining the EU's legal framework for secondary raw materials, this study seeks to identify successful policies, measures, and practices that have contributed to the EU's transition toward a circular economy.

In light of Vietnam's rapid economic growth, urbanization, and increasing resource demands, the need for sustainable resource management has become more urgent than ever. Vietnam's development trajectory provides a unique opportunity to embrace circular economy principles and leverage the EU's experiences and best practices. By adopting elements of the EU's legal framework and tailoring them to Vietnam's specific context, the country can pave the way for a more resilient, resource-efficient, and environmentally friendly future.

This paper aims to comprehensively analyze the EU's legal framework for secondary raw materials and present tailored recommendations to support Vietnam's transition toward sustainable resource management and the circular economy. Through this endeavor, we seek to contribute to the global effort of building a more sustainable and prosperous future for all.

2 Literature Review

2.1 Evolution of the Circular Economy Within the EU

The circular economy paradigm, while not new, has gained renewed interest as a strategic response to environmental sustainability and resource scarcity challenges. The European Union has been pivotal in advocating for this transition, with policy frameworks such as the EU Circular Economy Package setting the stage. The Waste Framework Directive, for instance, has been instrumental in steering member countries toward recycling and recovery targets. Scholarly works by authors like Geissdoerfer et al. and Kirchherr et al. provide comprehensive reviews of the circular economy's conceptual development and the EU's policy interventions.

2.2 Secondary Raw Materials: From Ecoslag to Policy

The transition to secondary raw materials usage is at the heart of the circular economy. Literature illustrates that these materials, often waste byproducts, can significantly reduce the environmental footprint when compared to primary resources. Case studies such as the application of Ecoslag in roadway construction demonstrate the practical benefits and feasibility of secondary raw materials. This body of research also explores the policy implications of embracing secondary materials, emphasizing the need for regulatory frameworks that facilitate their integration into the market.

2.3 Comparative Analyses of Circular Economy Adoption

Analyses of circular economy adoption reveal diverse trajectories among EU nations. The variability in recycling rates and policy implementation effectiveness provides insights into tailored strategies that can serve countries at different development stages. Comparative studies bring to light the potential for nations like Vietnam to leapfrog development stages by learning from the EU experience.

2.4 Systems Thinking in Circular Economy Frameworks

Systems thinking is crucial for understanding the interlinked nature of circular economies. Theoretical models like cradle-to-cradle propose a regenerative cycle where waste is minimized. The role of industrial ecology in promoting symbiotic industrial processes where one's waste becomes another's input is also highlighted in this literature.

2.5 Transition Theories in Circular Economy Policy

Transition theories such as the Multi-Level Perspective offer a lens through which the shift to a circular economy can be understood. Authors like Geels discuss the interplay between niche innovations, industry regimes, and overarching societal trends. This framework has been used to evaluate the adoption of circular economy practices and the role of policy in facilitating such transitions.

2.6 Innovations and Best Practices

Innovation is key in operationalizing the circular economy. Literature on this topic covers a wide range of initiatives, from municipal recycling programs to industrial symbiosis projects. Notably, the role of technological advancements in waste sorting and materials recovery is critical. These studies often discuss the policy measures that can stimulate and scale such innovations.

2.7 Case Studies: Ecoslag and Beyond

Case studies provide concrete examples of circular economy principles in action. For instance, the utilization of Ecoslag demonstrates how material recovery from waste can be successfully integrated into new products, contributing to both environmental sustainability and economic value.

2.8 Recent Findings and Forward Look

Emerging research points toward an increasing focus on digital technologies, such as blockchain and the Internet of Things (IoT), to enhance traceability and efficiency in circular systems. As the circular economy continues to evolve, policy implications include the need for digital infrastructure and data governance frameworks that support transparency and accountability in resource use.

2.9 Contribution of the Study

This study makes several distinct contributions to the existing body of knowledge on the circular economy, specifically in the context of Vietnam's development and alignment with EU standards.

2.10 Advancing Circular Economy Understanding

Firstly, it enhances the understanding of the circular economy by bridging the gap between theory and practice. By analyzing EU directives and their implementation success, this research provides a detailed road map for countries like Vietnam that are in the early stages of circular economy adoption.

2.11 Policy Development Framework

Secondly, it offers a nuanced framework for policy development, particularly in transitioning economies. The tailored recommendations for Vietnam, grounded in comparative analysis, serve as a valuable guide for policymakers seeking to reform waste management and promote secondary raw materials.

2.12 Quantitative Benchmarking

The study also contributes quantitative benchmarks on European recycling rates, providing a reference point for Vietnam's progress. By presenting detailed statistics and trend analyses, this research contributes to a deeper empirical understanding of waste management success and its drivers.

2.13 Methodological Innovation

Methodologically, this research pioneers the application of a combined approach, utilizing both theoretical frameworks and practical assessment tools, to evaluate the circular economy's implementation. This innovative approach can be adopted by other researchers and practitioners to assess circularity in different contexts.

2.14 Cross-Cultural Insights

Furthermore, by presenting Ecoslag as a case study, this study contributes cross-cultural insights into the applicability of circular economy principles. It highlights the potential for cross-border learning and the adoption of best practices from the EU in Vietnamese settings.

2.15 Enhancement of Sustainability Discourse

Lastly, by discussing the policy implications of the circular economy's adoption, this research adds to the global discourse on sustainability. It provides a solid foundation for discussions on how transitioning economies can balance economic development with environmental conservation.

This study's contributions are aimed at supporting sustainable development goals (SDGs) by informing policy, enhancing environmental protection, and promoting economic resilience through the adoption of circular economy practices.

3 EU's Legal Framework for Secondary Raw Materials

The European Union (EU) has established a comprehensive legal framework for secondary raw materials to promote sustainable resource management and advance the principles of a circular economy. The framework aims to optimize the use of secondary raw materials, minimize waste generation, and foster resource efficiency. This section provides an in-depth analysis of key EU directives, regulations, and policies that underpin the EU's approach to sustainable resource management through the promotion of secondary raw materials.

3.1 Waste Framework Directive (2008/98/EC)

The Waste Framework Directive serves as the cornerstone of the EU's legal framework for sustainable resource management. It lays down fundamental principles for waste management, including the promotion of waste prevention, reuse, recycling, and recovery. The directive encourages Member States to set ambitious recycling targets and implement measures to reduce waste generation. By prioritizing waste prevention and resource recovery, the Waste Framework Directive lays the foundation for the transition to a circular economy.[3]

3.2 Packaging and Packaging Waste Directive (94/62/EC)

The Packaging and Packaging Waste Directive focuses specifically on promoting the sustainable use of packaging materials. It sets targets for the recycling and recovery of packaging waste and encourages eco-design principles to improve the recyclability and environmental performance of packaging. This directive aims to reduce the environmental impact of packaging waste and promote the use of recycled materials in packaging production. It establishes extended producer responsibility schemes, where producers are responsible for financing the collection and recycling of packaging waste.[4]

---

[3] European Commission (2008).

[4] European Commission (1994).

3.3 Landfill Directive (1999/31/EC)

The Landfill Directive is another key element of the EU's legal framework for sustainable resource management. It establishes stringent requirements for the landfilling of waste to minimize environmental pollution and the release of greenhouse gases. The directive encourages the diversion of biodegradable waste from landfills, promoting waste prevention, recycling, and composting as more environmentally sound alternatives. By limiting the landfilling of waste, the Landfill Directive contributes to the reduction of waste and the promotion of circular economy practices.[5]

3.4 Circular Economy Package (2018)

The Circular Economy Package, adopted in 2018, is a comprehensive set of initiatives to further advance the circular economy in the EU. It includes revised waste legislation, setting higher recycling targets and reducing landfilling. The package also introduces measures to promote eco-design, foster product durability, and enhance consumer information on sustainable products. Furthermore, it aims to address specific waste streams, such as plastics, by setting targets for recycling and reducing marine litter. The Circular Economy Package reinforces the EU's commitment to promoting sustainable resource management through circular economy principles.[6]

The EU's legal framework for secondary raw materials demonstrates a holistic approach to sustainable resource management and circular economy practices. By prioritizing waste prevention, recycling, and the responsible management of end-of-life products, the EU aims to create a more resource-efficient and environmentally sustainable economy.

Through the comprehensive legal framework, the EU has successfully enhanced resource recovery, reduced waste generation, and promoted the use of secondary raw materials in various sectors. The implementation of extended producer responsibility schemes has incentivized producers to design products with circularity in mind, fostering a more sustainable product life cycle.

In conclusion, the EU's legal framework for secondary raw materials provides valuable insights and successful practices for promoting sustainable resource management and circular economy principles. By aligning their policies and practices with the EU's legal framework, countries like Vietnam can optimize resource use, minimize waste, and contribute to the global effort toward a more sustainable and circular economy. The next section will delve into a comparative analysis of Vietnam's regulations and the EU's legal framework to identify areas of potential

[5] European Commission (1999).

[6] Scharff (2018).

alignment and inform tailored recommendations for Vietnam's sustainable resource management journey.

4 Theoretical Framework for Circular Economy Implementation

4.1 Systems Thinking

1. Introduction to Systems Thinking: Systems thinking is a fundamental concept in transitioning to a circular economy. It involves viewing the economy as an interconnected and interdependent system, where products, materials, and resources circulate in closed loops. This approach contrasts with the traditional linear model, emphasizing the importance of feedback loops, holistic understanding, and the relationships between different components.
2. Circular Systems Models: Explore theoretical models that depict how circular systems operate. These models can include the cradle-to-cradle approach, industrial ecology, and biomimicry, showcasing how these frameworks promote resource efficiency, reduced waste, and sustainable production.
3. Resource Loops and Feedback Mechanisms: Discuss the key components of circular systems, such as resource loops and feedback mechanisms. Explain how materials are continuously cycled, remanufactured, and repurposed within a circular economy. Emphasize how feedback mechanisms enable continuous improvement and adaptation.
4. Interdependence in Circular Systems: Highlight the interdependence of various elements within a circular economy. This includes the role of businesses, consumers, government policies, and technological advancements in creating a functioning circular system. Show how each element affects and is affected by others in a circular economy.

4.2 Transition Theories

1. Multi-Level Perspective (MLP): Introduce the Multi-Level Perspective as a theoretical framework for understanding the transition from a linear to a circular economy. Explain the MLP's key components, including landscape, regime, and niche, and how these interact to facilitate or hinder the transition. Use real-world examples to illustrate the application of MLP in circular economy transitions.
2. Sustainability Transition Framework: Discuss the Sustainability Transition Framework, which focuses on the dynamics of societal change toward sustainability. Show how this framework can guide the transformation process in various domains, including production and consumption patterns, energy use, and waste management.

3. Innovation Theories: Explore theories related to innovation, as innovation plays a vital role in transitioning to a circular economy. Discuss concepts such as disruptive innovation, diffusion of innovations, and open innovation models. Explain how these theories can facilitate the adoption of circular practices and technologies.

4.3 Policy Theories

1. Extended Producer Responsibility (EPR): Examine the theory and application of Extended Producer Responsibility in the context of circular economy policy. Describe how EPR places the responsibility on producers for the entire lifecycle of their products, encouraging eco-design, recycling, and responsible disposal. Provide case studies to illustrate the effectiveness of EPR policies.
2. Eco-Design Theory: Explore the theoretical foundations of eco-design, emphasizing the integration of environmental considerations into product design. Discuss how eco-design principles, such as designing for disassembly and recyclability, are crucial in achieving circularity. Present examples of eco-design in practice and its positive impacts.
3. Green Public Procurement (GPP): Explain the theoretical underpinnings of Green Public Procurement and how it influences circular economy policies. Highlight how GPP encourages government agencies to prioritize environmentally friendly products and services, stimulating demand for circular products. Showcase successful instances of GPP implementation.

By delving into these theoretical frameworks, the article aims to provide a solid foundation for understanding the principles and strategies that underpin the transition to a circular economy. It illustrates the importance of systems thinking, transition theories, and policy theories in guiding and facilitating the shift toward sustainable and circular resource management.

5 Methodology for Circular Economy Transition

Transitioning to a circular economy, particularly with an emphasis on secondary raw materials, requires a structured approach to measure progress and inform strategic decisions. The methodology outlined below utilizes specific assessment tools designed for this purpose:

5.1 Circular Economy Performance Dashboard

Purpose and Application: This dashboard serves as an all-encompassing tool for organizations across various sectors to gauge their level of circularity. It tracks critical indicators such as the ratio of recycled content in products and waste reduction metrics.

Implementation Steps:

1. Establish baseline metrics for current resource use and waste production.
2. Regularly monitor and update data on secondary material usage and waste reduction efforts.
3. Analyze trends to identify improvement opportunities and implement targeted actions.

5.2 Circular Transition Indicators

Objective: These indicators are tailored to assess how effectively an organization or sector is transitioning from a linear model to a circular one, with a specific focus on the incorporation of secondary raw materials.

Key Metrics:

1. The percentage of secondary materials in product composition.
2. The volume of material redirected from landfills to recycling or reuse.
3. The expansion or establishment of recycling and remanufacturing facilities.

5.3 Life Cycle Assessment (LCA)

Purpose: LCA is a critical tool for evaluating the environmental footprint of products or processes, comparing the benefits of using recycled materials against those of virgin resources.

Approach:

1. Conduct a comprehensive assessment of the product lifecycle, from raw material extraction to end-of-life.
2. Calculate the environmental savings in terms of greenhouse gas emissions, energy use, and other impacts from integrating secondary raw materials.
3. Use findings to optimize processes for greater environmental efficiency.

5.4 Circular Economy Maturity Models

Function: These models offer a framework to evaluate an organization's maturity level regarding circular economy practices, including the adoption of secondary raw materials.

Evaluation Process:

1. Determine the current stage of circularity within the organization.
2. Set specific, measurable targets for enhancing circular practices.
3. Monitor progress against these targets, adjusting strategies as necessary to advance through maturity levels.

5.5 Supply Chain Analysis

Significance: Understanding the supply chain dynamics for secondary raw materials is essential to ensure their sustainable and resilient supply.

Analytical Framework:

1. Map out the supply chain for key secondary materials.
2. Identify potential risks and opportunities related to material sourcing.
3. Develop strategies to mitigate risks and leverage opportunities, enhancing supply chain sustainability.

5.6 Environmental and Economic Impact Assessment

Goal: To understand the broader impacts of transitioning to circular practices, specifically the use of secondary raw materials, on both environmental outcomes and economic performance.

Assessment Techniques:

1. Quantify cost savings and resource efficiency gains from adopting circular practices.
2. Evaluate the environmental benefits, such as reduced carbon footprint and lower resource depletion rates.
3. Translate these benefits into compelling business cases to support investment in circular economy initiatives.

By adopting and adapting these tools, organizations can gain a comprehensive understanding of their circular economy transition, enabling data-driven decisions that foster continuous improvement. This methodical approach supports the effective management of secondary raw materials, aligning operational practices with sustainability goals for a resilient, circular future.

6 Data and Empirical Analysis

To comprehensively assess the EU's success in implementing its legal framework and the current state of resource management in Vietnam, we'll delve into various forms of data, both quantitative and qualitative.

6.1 Recycling Rates and Waste Generation in the EU

We will provide detailed statistical data on recycling rates in various EU member states, highlighting the significant progress made in recent years. This data will illustrate the effectiveness of recycling programs and regulations within the EU, emphasizing the reduction of waste generation and the impact on resource utilization. We can include tables and charts to present this data clearly.

Objective: To present detailed statistical data on recycling rates in various EU member states, showcasing recent progress and illustrating the effectiveness of recycling programs and regulations.

Methodology: Utilize tables and charts to visually represent recycling rates, emphasizing the reduction in waste generation and its impact on resource utilization.

6.2 Environmental Impacts in the EU

This section will explore the environmental impacts of the EU's resource management practices. We can discuss reductions in greenhouse gas emissions, energy consumption, and overall ecological footprint, all of which can be supported by relevant data and case studies. This empirical evidence will underscore the positive environmental outcomes resulting from a circular economy approach.

Objective: Explore the environmental impacts of the EU's resource management practices, focusing on reductions in greenhouse gas emissions, energy consumption, and overall ecological footprint.

Methodology: Support findings with relevant data and case studies, emphasizing the positive environmental outcomes resulting from the adoption of a circular economy approach.

6.3 Vietnam's Current Resource Management Scenario

To provide a comparative analysis, we will present quantitative data on resource management in Vietnam. This could encompass data on waste generation, recycling rates, and the environmental consequences of the current linear economic model.

Interviews and surveys conducted with key stakeholders and government officials in Vietnam will provide insights into the country's resource management challenges and opportunities.

Objective: Provide a comparative analysis by presenting quantitative data on resource management in Vietnam, including waste generation, recycling rates, and the environmental consequences of the current linear economic model.

Methodology: Conduct interviews and surveys with key stakeholders and government officials in Vietnam to gather insights into the country's resource management challenges and opportunities.

In examining the landscape of resource management, the European Union (EU) has emerged as a global leader in adopting a circular economy model, with a focus on utilizing secondary raw materials efficiently. The success of the EU's recycling initiatives is evident when delving into statistical data on recycling rates across its member states. Visual aids, such as tables and charts, will be instrumental in illustrating the remarkable progress witnessed in recent years.

The statistical analysis of recycling rates in EU member states reveals a positive trajectory, showcasing a substantial increase in the proportion of recycled materials. This progress is particularly noteworthy as it highlights the tangible impact of recycling programs and regulations within the EU. The incorporation of visual aids not only enhances the clarity of the data but also serves to underscore the remarkable achievements in waste reduction and resource utilization.

Furthermore, the environmental impacts stemming from the EU's circular economy practices are significant and multi-faceted. Reductions in greenhouse gas emissions and energy consumption stand out as key indicators of the positive ecological outcomes resulting from the circular approach. Case studies will be employed to provide concrete examples supporting these reductions, demonstrating the holistic benefits that extend beyond waste management.

The circular economy practices adopted by the EU contribute not only to waste reduction but also to an overall decrease in its ecological footprint. The emphasis on reusing materials and minimizing waste generation has led to a more sustainable and environmentally friendly resource management approach. This holistic impact aligns with the broader global goals of environmental conservation and sustainable development.

Shifting the focus to Vietnam, a comparative analysis of its current resource management scenario provides valuable insights. Presenting quantitative data on waste generation and recycling rates in Vietnam will offer a clear picture of the existing challenges and opportunities. Additionally, qualitative insights from interviews and surveys with key stakeholders and government officials will provide a nuanced understanding of the country's unique circumstances.

The comparative analysis between the EU and Vietnam will shed light on potential disparages and areas for improvement. Understanding the disparities in waste management practices will enable policymakers and stakeholders in Vietnam to identify strategic areas for intervention and development. By drawing upon the EU's successful circular economy model, Vietnam can leverage this comparative analysis to shape its own resource management policies effectively.

In conclusion, this comprehensive analysis combines quantitative data and qualitative insights to present a clear understanding of the EU's success in resource management and the current scenario in Vietnam. The EU's achievements in recycling and its positive environmental impacts serve as valuable lessons for nations seeking to transition toward more sustainable and circular resource management practices. The comparative analysis with Vietnam serves as a roadmap for identifying opportunities and formulating effective policies to enhance resource efficiency and environmental sustainability.

6.4 Case Studies in Vietnam: The Ecoslag Solution

6.4.1 Introduction

Ecoslag emerges as a high-performance material for roadway construction, offering an eco-friendly alternative to traditional aggregates. Developed by Green Materials Company (Công ty TNHH Vật Liệu Xanh), this innovative material not only enhances the structural integrity of roadways but also contributes to environmental sustainability.

6.4.2 Product Description

Ecoslag is a type of eco-friendly aggregate designed for superior roadway infrastructure. It is produced from recycled materials from the steel manufacturing process, aligning with the global movement toward sustainable development.

6.4.3 Technical Specifications

The Ecoslag exhibits superior technical qualities such as:

- Particle Size Distribution: Ranging from fine to coarse particles, it offers a flexible composition for various construction needs.
- Specific Gravity: Ecoslag has a higher specific gravity compared to conventional stone, implying a denser and potentially more durable construction material.
- Water Absorption: It boasts lower water absorption rates, indicative of its resilience against weathering and degradation.

6.4.4 Advantages Over Conventional Materials

Ecoslag is distinguished by its:

- Higher durability and load-bearing capacity compared to traditional stone materials.
- Reduced environmental footprint, utilizing recycled materials and reducing the need for quarrying natural stone.

Applications in Road Construction

It has been effectively used in "Open Porous Asphalt" (OPA), providing a road surface that offers better drainage and a reduced risk of hydroplaning due to its permeable nature. This demonstrates Ecoslag's potential to revolutionize road infrastructure by improving safety and longevity.

Environmental Impact

Ecoslag stands as a testament to eco-friendly construction practices, where its use in roadways contributes to a significant reduction in carbon emissions and conserves natural resources.

Conclusion

This case study concludes that Ecoslag, provided by Green Materials Company, is a forward-thinking solution that supports Vietnam's commitment to sustainable infrastructure and environmental stewardship. Its application in road construction showcases the viability of recycled materials in major infrastructural projects, paving the way for broader adoption of green construction practices.

To craft a detailed and accurate chapter, you would integrate specific quantitative data, as well as microscopic images that reveal the material's unique properties, as provided in the visual documentation. Additionally, direct quotes or paraphrased material from reputable sources would support claims about Ecoslag's performance and environmental benefits.

7 Case Studies and Best Practices in Circular Economy and Sustainable Resource Management

7.1 IKEA's Sustainable Approach to Recycled Materials

IKEA, a global leader in home furnishings, has been at the forefront of sustainable practices, emphasizing the use of recycled materials to create high-quality products while simultaneously reducing production costs. Their commitment to sustainability is evident in their ambitious goal to source renewable or recycled materials exclusively by 2030.[7] This vision is underpinned by their dedication to providing innovative solutions that extend the lifespan of products and materials, contributing to a circular economy.

7.2 *Recycling Metal Responsibly*

One of the key materials in IKEA's product range is metal, appreciated for its strength, durability, and affordability. What sets IKEA apart is their commitment to making metal even more sustainable by ensuring responsible sourcing and production. By adhering to strict sustainability criteria, they aim to reduce the environmental impact of metal extraction and processing. IKEA recognizes that sourcing metal responsibly is not only beneficial for the environment but also crucial for delivering products that stand the test of time.

7.3 *Bamboo As a Sustainable Material*

Bamboo is a remarkable material that combines strength, sustainability, and versatility. IKEA has embraced the potential of bamboo and is increasingly using it in the creation of durable and aesthetically pleasing products. By harnessing the strength of bamboo, IKEA contributes to the reduction of resource-intensive materials while providing customers with sustainable alternatives. The choice to incorporate bamboo into their product range aligns with their commitment to environmental responsibility.

7.3.1 The Shift Toward Recycled or Renewable-Based Plastic

IKEA's ambition extends to their plastic usage as well. They aspire to ensure that all plastic used in their products is derived from renewable or recycled sources. This commitment begins with the elimination of single-use plastics and the adoption of various sustainable plastic materials. By making this transition, IKEA takes substantial steps toward reducing plastic waste and its environmental impact. This approach aligns with the broader global movement to mitigate the harmful effects of plastic pollution and underscores IKEA's responsibility as a major player in the industry.

---

[7] IKEA (2020).

7.4 Sustainable Wood Sourcing

Wood is a cornerstone of IKEA's product portfolio, and their commitment to responsible wood sourcing is paramount. Through their partnership with the Forest Stewardship Council (FSC), IKEA is actively involved in promoting responsible forest management. This partnership ensures that forests are managed in a way that preserves biodiversity and safeguards the rights of local communities. The goal is to maintain the world's forests for future generations while meeting the demands for wood as a raw material.

IKEA's dedication to sustainable wood sourcing has led to a significant achievement: all the wood they use is either FSC-certified or sourced from recycled materials. This commitment to stringent standards not only assures customers of the responsible origins of IKEA's products but also sets an industry-leading example for responsible wood sourcing.

In conclusion, IKEA's approach to using recycled materials and sustainable sourcing practices is a testament to their commitment to environmental responsibility. Their ambitious goal to rely exclusively on renewable or recycled materials by 2030 reflects a dedication to sustainability that extends to every aspect of their business. IKEA's leadership in embracing materials like bamboo, sustainable plastics, and responsibly sourced wood sets a high standard for the industry and serves as an inspiration for others looking to make a positive impact on the environment.

Many companies are actively using recyclable materials or secondary raw materials in their products and packaging to reduce environmental impact and promote sustainability. Here are some notable examples:

1. Nike: Nike has been using recycled polyester from plastic bottles in their athletic wear for years. They also collect and recycle old athletic shoes into material for sports surfaces and new shoes.
2. Coca-Cola: Coca-Cola has made a commitment to use 50% recycled content in its PET plastic bottles by 2030. They also support recycling initiatives and have introduced PlantBottle packaging made partially from sugarcane.
3. Apple: Apple has been working to use recycled and renewable materials in its products. For instance, they use recycled aluminum in some of their devices.
4. Adidas: Adidas has introduced sneakers made from recycled ocean plastic. They are also actively working to reduce waste and use more sustainable materials in their products.
5. Unilever: Unilever, the consumer goods giant, is committed to making all of its plastic packaging recyclable, reusable, or compostable by 2025. They are also working to use more recycled plastic in their packaging.
6. HP: HP has made efforts to use recycled plastic in its printer cartridges and to reduce plastic in its packaging.
7. Procter & Gamble: P&G, the maker of many household products, has set a goal to use 100% recyclable or reusable packaging by 2030 and to reduce its use of virgin petroleum plastic by 50%.

8. Dell: Dell has a closed-loop recycling program where they use plastics from old electronics to make new ones.
9. Nestlé: Nestlé is working to achieve 100% recyclable or reusable packaging by 2025 and to reduce its use of virgin plastics.
10. LEGO: LEGO is investing in sustainable materials and aims to make all of its products and packaging from sustainable materials by 2030.

These companies are part of a broader movement in the business world to reduce waste and use resources more sustainably. They are actively incorporating recyclable materials and secondary raw materials into their products and packaging to minimize their environmental footprint.

8 Comparative Analysis of Vietnam's Regulations

The comparative analysis of Vietnam's regulations and the EU's legal framework for secondary raw materials sheds light on the similarities, differences, challenges, and opportunities that Vietnam faces in transitioning toward sustainable resource management and circular economy practices. This section critically examines Vietnam's current regulations and identifies areas where alignment with the EU's best practices can accelerate progress toward a more sustainable and resource-efficient economy.

8.1 Current Legal Framework for Secondary Raw Materials in Vietnam

Vietnam has made strides in environmental regulation and resource management, but its current legal framework for secondary raw materials is still evolving. The country has implemented various laws and policies related to waste management and recycling, aiming to reduce environmental pollution and foster resource recovery. Key regulations include the Law on Environmental Protection (Law No. 72/2020/QH14), which establish the legal basis for waste prevention, treatment, and recycling. Vietnam has also introduced measures to promote renewable energy and sustainable production practices, addressing broader environmental concerns.

8.2 Challenges in Waste Management and Resource Recovery

Despite progress, Vietnam faces several challenges in waste management and resource recovery. Insufficient waste infrastructure, limited waste collection systems, and lack of recycling facilities hinder effective resource recovery from waste.

Informal waste practices and a lack of awareness among the public and industry stakeholders contribute to low recycling rates. The informal sector's involvement in waste management raises concerns about health and environmental risks. Additionally, limited financial and technical resources pose obstacles to advancing sustainable resource management practices in Vietnam.

8.3 Opportunities for Aligning with the EU's Legal Framework

Despite challenges, Vietnam can leverage the EU's legal framework to capitalize on best practices and foster sustainable resource management. The EU's emphasis on waste prevention, extended producer responsibility, and eco-design aligns with Vietnam's vision of moving toward a circular economy. By adopting and adapting elements of the EU's legal framework, Vietnam can bolster its resource management strategies and accelerate the transition to a more sustainable economy.

8.4 Extended Producer Responsibility (EPR) Implementation

The EU's success in implementing EPR schemes can serve as a model for Vietnam. By incorporating EPR into its legal framework, Vietnam can encourage producers to take greater responsibility for their products' end-of-life management. This measure would drive eco-design and promote the design of products with recyclability and resource efficiency in mind.

8.5 Strengthening Waste Collection and Recycling Infrastructure

Vietnam can draw inspiration from the EU's efforts to improve waste collection and recycling infrastructure. Investment in modern waste treatment facilities and expanding waste collection networks would facilitate greater resource recovery. Additionally, establishing a formalized recycling industry and providing incentives for waste recycling businesses would create economic opportunities while reducing environmental burden.

8.6 *Fostering Public Awareness and Behavior Change*

The EU's focus on fostering public awareness and promoting behavioral change toward sustainable consumption and waste management is essential for Vietnam. Raising awareness about the value of secondary raw materials and promoting responsible waste disposal practices can drive public participation and increase recycling rates.

8.7 *Collaboration and Capacity Building*

Collaboration between government agencies, industries, academia, and civil society is crucial for Vietnam's successful transition to sustainable resource management. The EU's experience in fostering collaboration and knowledge exchange can inform Vietnam's efforts to build capacity and implement circular economy practices effectively.

In conclusion, the comparative analysis highlights the opportunities for Vietnam to align its legal framework with the EU's best practices for sustainable resource management and circular economy principles. While facing challenges, Vietnam can benefit from the EU's expertise and experiences in promoting secondary raw materials, waste prevention, and resource efficiency. By embracing and adapting elements of the EU's legal framework, Vietnam can chart a path toward a more sustainable and resilient future, driving economic growth while preserving its natural resources for future generations. The next section will provide specific recommendations tailored to Vietnam's context to accelerate progress toward sustainable resource management and circular economy practices.

Based on the comparative analysis of Vietnam's regulations and the EU's legal framework for secondary raw materials, the following recommendations are tailored to support Vietnam's transition toward sustainable resource management and circular economy practices.

9 Vietnam's Challenges and Limitations in Studying According to EU Law

Implementing the EU's legal framework for secondary raw materials in Vietnam presents several challenges and limitations, which must be addressed for a successful transition to a circular economy. These challenges encompass financial constraints, cultural differences, and regulatory hurdles:

9.1 Financial Constraints

Investment: Transitioning to a circular economy often requires significant investment in new technologies, infrastructure, and research and development. Vietnam may face financial constraints in making these investments, especially in the early stages of adopting circular practices.

Cost of Transition: Shifting from a linear to a circular model may initially increase costs, as new processes and technologies are integrated. It can be challenging for businesses, particularly small and medium-sized enterprises (SMEs), to absorb these costs without financial support.

9.2 Cultural Differences

Consumer Behavior: Vietnamese consumer behavior and preferences may differ from those in the EU. Adapting to a circular economy requires changes in consumption patterns and a willingness to embrace products made from secondary raw materials. Changing deeply ingrained consumer habits can be a cultural challenge.

Business Practices: Companies in Vietnam may have traditional business practices and supply chain models that are not aligned with circular principles. Implementing circular practices may require significant changes in corporate culture and business strategies.

9.3 Regulatory Hurdles

Legal Framework: While adopting the EU's legal framework can provide valuable guidance, Vietnam has its unique legal and regulatory environment. Adapting EU regulations to Vietnam's context and ensuring their effective implementation can be a complex process.

Enforcement: Regulatory enforcement and monitoring may be challenging, as it requires the commitment and capacity of government agencies to ensure compliance with circular economy regulations.

9.4 Infrastructure and Technology

Waste Infrastructure: Vietnam may lack the necessary waste collection, sorting, and recycling infrastructure to support efficient secondary raw material recovery. Developing this infrastructure can be resource-intensive and time-consuming.

Technology Access: Access to advanced recycling and remanufacturing technologies is crucial for effectively utilizing secondary raw materials. Ensuring that these technologies are accessible and affordable in Vietnam is a significant challenge.

9.5 *Public Awareness*

Education: Building public awareness about the benefits of secondary raw materials and the circular economy is an ongoing challenge. It involves educational campaigns and outreach efforts to change public perceptions and behaviors.

Resistance to Change: There may be resistance to change, as people are often reluctant to adopt new practices or products. Overcoming this resistance and fostering a culture of acceptance for circular economy principles is a significant limitation.

9.6 *Data and Monitoring*

Data Collection: Establishing reliable data collection and reporting systems for measuring progress in secondary raw material use and circular practices can be a challenge.

Monitoring Capacity: Government agencies and organizations may lack the capacity to effectively monitor and evaluate the impact of circular economy initiatives.

9.7 *Capacity Building*

Skills and Knowledge: Developing the necessary skills and knowledge among the workforce and government officials to implement circular practices can be a limitation. Training and capacity-building programs are required.

Addressing these challenges and limitations will be essential for Vietnam to effectively implement the EU's legal framework for secondary raw materials and successfully transition to a circular economy. It will require a concerted effort from the government, businesses, civil society, and the public to overcome these hurdles and embrace the circular economy principles.

10 Recommendations for Vietnam

10.1 Recommendations for Vietnam

1. Strengthening Legal and Regulatory Frameworks: Vietnam should enhance its legal and regulatory frameworks to support circular economy practices. This includes the adoption of legislation similar to the EU's Waste Framework Directive and Packaging and Packaging Waste Directive, focusing on waste reduction, recycling, and the use of secondary raw materials.
2. Extended Producer Responsibility (EPR) Programs: Implementing EPR programs would encourage manufacturers and producers to take responsibility for the entire lifecycle of their products, including end-of-life management. This policy has been effective in the EU for improving recycling rates and can stimulate innovation in product design for recyclability and durability in Vietnam.
3. Investment in Recycling Infrastructure: Significant investments are needed to develop and upgrade recycling facilities and infrastructure in Vietnam. This will enhance the country's capacity to process secondary raw materials locally, reducing reliance on imports and fostering a domestic circular economy.
4. Public Awareness and Education Campaigns: To foster a culture of recycling and sustainability, it is crucial to launch comprehensive public awareness and education campaigns. These should highlight the importance of circular economy practices and how individuals and businesses can contribute to resource efficiency.
5. Research and Development (R&D) in Circular Technologies: Encouraging R&D in circular technologies and sustainable material alternatives can drive innovation in Vietnam. This includes support for startups and businesses developing products and processes that are aligned with circular economy principles.
6. Public–Private Partnerships (PPPs): Facilitate PPPs to leverage the strengths and resources of both the public and private sectors. These partnerships can accelerate the implementation of circular economy projects, from waste management solutions to the development of sustainable products.

10.2 Policy Implications

- Economic Growth and Competitiveness: Transitioning toward a circular economy can stimulate economic growth by creating new business opportunities and jobs in recycling, sustainable materials, and green technologies. It also positions Vietnam as a competitive player in the global market for sustainable products and services.
- Environmental Protection: By reducing waste and promoting the use of secondary raw materials, Vietnam can significantly lower its environmental footprint, contributing to global efforts to combat climate change and preserve biodiversity.

- Resource Security: Implementing circular economy practices enhances Vietnam's resource security by minimizing dependency on imported raw materials and mitigating the impacts of global supply chain disruptions.
- Social Well-being: A circular economy supports social well-being by creating employment opportunities in green industries and improving the quality of the environment for communities.
- International Cooperation: Vietnam's adoption of circular economy principles can strengthen its international partnerships, especially with the EU, by aligning its policies with global sustainability goals and standards.
- In conclusion, by embracing these recommendations and understanding their policy implications, Vietnam can make significant strides toward sustainable resource management and a circular economy. This not only benefits the environment but also fosters economic resilience, social inclusivity, and a sustainable future for generations to come.

11 Detailed Statistical Data on European Recycling Rates

Waste recycling in the European Union (EU) has been on the rise. The EU's municipal waste recycling rate has been gradually improving over the past few years and has now reached 49%. Nevertheless, 24% of municipal waste is still being landfilled. Recycling rates vary across different waste streams, ranging from 64% for total packaging waste to 39% for electrical and electronic waste.[8]

Our waste-related indicators also provide the following insights:

Waste generation (excluding mineral wastes) in Europe increased by 1% from 2010 to 2020.[9]

On average, each European citizen generates approximately 4.8 tons of waste annually, with about 39% of it being recycled. Recycling rates exhibit significant variations by country. Eight countries have achieved recycling rates of 50% or higher, while another eight countries recycle less than 20% of their municipal waste.[10]

Overall, the proportion of waste sent to landfills in the EU-27 decreased from 23% to 16% between 2010 and 2020, even though the total amount of waste generated has continued to grow.[11]

The EU generates massive amounts of waste from manufacturing, food, construction, electronics, textiles, and more is being generated each year.

In addition to losing valuable resources, waste can have significant negative impacts on human health and the environment. For example, incinerating waste

[8] WasteSMART/EEA (2023).

[9] WasteSMART/EEA (2023).

[10] WasteSMART/EEA (2023).

[11] WasteSMART/EEA (2023).

can contribute to air pollution, landfills might contaminate water and land. That is why the EU developed a waste policy that keeps products in use for as long as possible and sets strict standards for how to responsibly dispose of them when they can no longer be used. Policies and initiatives also focus on preventing waste generation in the first place by, for example, reducing food waste and making products easier to repair.

Many reuse and recycling initiatives can contribute to reducing the loss of resources in the EU:

- Reuse centers and second-hand shops make old products available to new users.
- Biowaste is turned into biogas and fertilizer.
- Secondary materials are created from building waste.
- Metals in electronic waste, such as from computers, gadgets, and batteries, are recovered and can be used in new devices.

The EU generates massive amounts of waste from manufacturing, food, construction, electronics, textiles... And more is being generated each year.

In addition to losing valuable resources, waste can have significant negative impacts on human health and the environment. For example, incinerating waste can contribute to air pollution, landfills might contaminate water and land.

That is why the EU developed a waste policy that keeps products in use for as long as possible and sets strict standards for how to responsibly dispose of them when they can no longer be used. Policies and initiatives also focus on preventing waste generation in the first place by, for example, reducing food waste and making products easier to repair.

Many reuse and recycling initiatives can contribute to reducing the loss of resources in the EU (see Fig. 1):

- Reuse centers and second-hand shops make old products available to new users.
- Biowaste is turned into biogas and fertilizer.
- Secondary materials are created from building waste.
- Metals in electronic waste, such as from computers, gadgets and batteries, are recovered and can be used in new devices.

The indicator assesses the proportion of recycled materials integrated back into the economy, thereby decreasing the need for primary raw material extraction. This indicator, known as the circularity rate, is calculated as the ratio of materials reused in a circular manner to the total material consumption. Total material consumption is determined by combining domestic material consumption (DMC) and the recycled material usage. DMC is a metric defined within the broader scope of economy-wide material flow accounts (Fig. 2).

To approximate the circular use of materials, we consider the amount of waste recycled in local recovery facilities, accounting for recovery operations R2 to R11 as outlined in the Waste Framework Directive 75/442/EEC. The volume of waste imported for recovery is subtracted, while exported waste intended for recovery abroad is added. The quantities of imported and exported waste slated for recycling are estimated using European trade statistics.

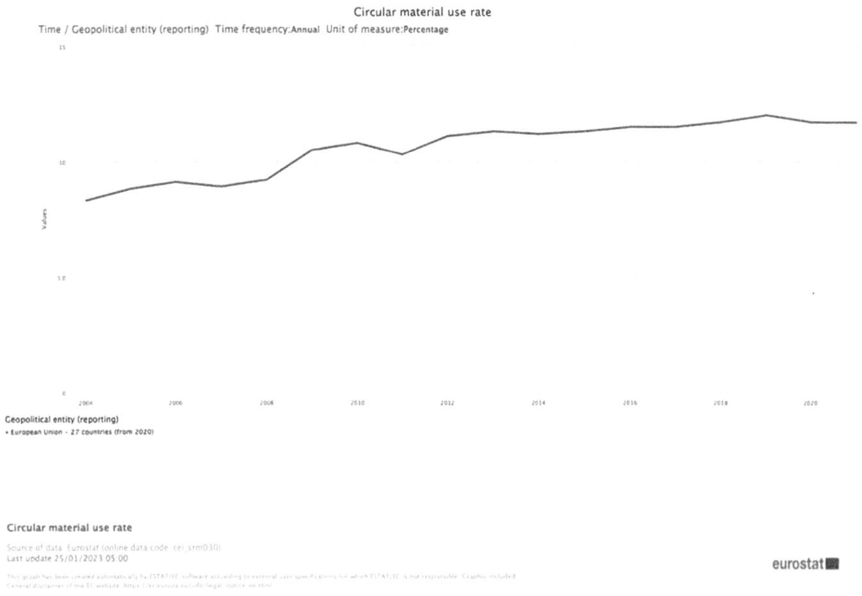

Fig. 1 Circular material use rate, 2014–2020 (Circular material use rate—Secondary raw materials—Circular economy indicators (line chart) https://ec.europa.eu/eurostat/databrowser/view/cei_srm030/default/bar?lang=en)

A higher circularity rate signifies a greater substitution of secondary materials for primary raw materials, thereby reducing the environmental impact associated with primary material extraction.

Furthermore, it's important to note that several countries within the EU demonstrated recycling rates that exceeded the EU's overall average. Notable among these nations are France, Italy, Estonia, Germany, and Austria. These countries consistently showcased a strong commitment to sustainable material management practices, with recycling rates higher than the EU average, reinforcing their dedication to environmental responsibility and circular economy principles.

12 Conclusion

In conclusion, the European Union's (EU) legal framework for secondary raw materials provides a robust foundation for sustainable resource management and circular economy practices. Through key directives, regulations, and policies, the EU prioritizes waste prevention, recycling, and resource efficiency to optimize the use of secondary raw materials and reduce environmental impacts. This legal framework serves as a valuable reference for countries like Vietnam, which are

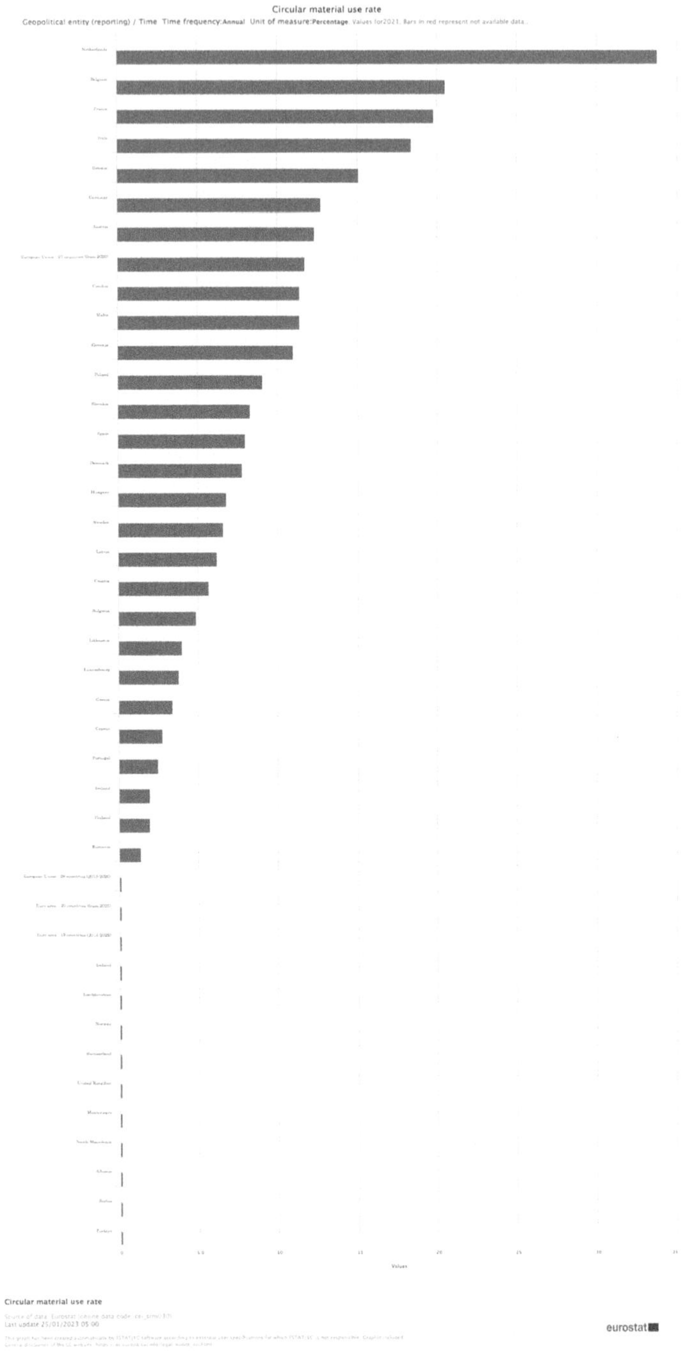

Fig. 2 Displays the circular material use rate across countries (Circular material use rate—Secondary raw materials—Circular economy indicators (bar chart) https://ec.europa.eu/eurostat/databrowser/view/cei_srm030/default/bar?lang=en)

seeking to embrace sustainable resource management practices and transition toward a circular economy.

The comparative analysis of Vietnam's regulations and the EU's legal framework has identified areas of potential alignment and revealed opportunities for Vietnam to accelerate progress toward sustainability. Vietnam faces challenges in waste management infrastructure, resource recovery, and public awareness, but these challenges can be overcome through targeted measures and learning from the EU's best practices.

The recommendations presented in this paper provide tailored guidance for Vietnam's sustainable resource management journey. By implementing extended producer responsibility schemes, strengthening waste collection and recycling infrastructure, fostering public awareness, and collaborating with the private sector, Vietnam can advance circular economy principles and optimize its resource use.

Furthermore, case studies and best practices from EU member states showcase successful circular economy initiatives in various sectors, offering valuable insights for Vietnam to adapt and implement similar approaches. By leveraging these examples and building strong partnerships between government, industry, academia, and civil society, Vietnam can accelerate the transition to a more sustainable and resilient economy.

The adoption of circular economy principles and sustainable resource management practices not only benefits the environment but also presents economic opportunities and social benefits. By reducing resource consumption, minimizing waste, and promoting eco-friendly production, Vietnam can enhance its economic competitiveness and build a more prosperous and inclusive society.

In the global effort to combat environmental challenges and achieve sustainable development goals, collaboration and knowledge exchange play a vital role. Vietnam can collaborate with the EU and other countries to exchange experiences, technologies, and best practices, fostering international cooperation for a more sustainable future.

In conclusion, the alignment of Vietnam's legal framework with the EU's approach to secondary raw materials and circular economy practices holds immense potential for sustainable development. By embracing circular economy principles, Vietnam can pave the way for a greener, more resource-efficient economy, ensuring the preservation of its natural resources for generations to come and contributing to the global transition toward a more sustainable and circular world.

13 Future Perspective

As we cast our gaze into the future, the adoption of the European Union's (EU) legal framework for secondary raw materials stands as a transformative move that can position Vietnam at the forefront of sustainable resource management and circular economy practices. In response to the escalating environmental and resource challenges posed by the traditional linear model of "take, make, dispose," the EU has emerged as a global leader in advocating for a circular economy. This model prioritizes reducing resource consumption, curbing waste generation, and mitigating ecological degradation.

Vietnam, by aligning itself with the EU's progressive approach, has the potential to embark on a journey of sustainable development that not only safeguards its natural resources but also contributes to global efforts in environmental conservation. The implementation of recommended measures, coupled with a commitment to fostering innovation and facilitating collaboration, can unlock economic opportunities for Vietnam while concurrently addressing pressing environmental concerns.

Embracing circular economy principles will undoubtedly lead to enhanced environmental protection. Vietnam can anticipate a reduction in waste generation, mitigating the adverse impacts of traditional linear models on ecosystems. The circular approach also promises increased resource efficiency, ensuring that the country maximizes the utility of its resources while minimizing their depletion.

In the context of a global shift toward circular practices, Vietnam's alignment with the EU's approach holds the key to fostering international cooperation. As circular economy principles gain momentum globally, Vietnam's commitment will contribute significantly to a more resilient and sustainable world. Collaborative efforts with the EU and other nations pursuing circularity will create synergies, amplifying the positive impact on a global scale.

The future perspective for Vietnam, therefore, is one characterized by growth, resilience, and an unwavering commitment to sustainable development through circularity. By adopting and adapting the EU's legal framework, Vietnam can become a beacon for other nations, showcasing the potential for harmonizing economic growth with environmental sustainability. As the world grapples with the challenges of resource depletion and environmental degradation, Vietnam can emerge not only as an economic success story but also as a model for responsible and sustainable resource management on the global stage.

References

European Commission. (1994). *European Parliament and Council Directive 94/62/EC of 20 December 1994 on packaging and packaging waste*. https://eur-lex.europa.eu/eli/dir/2008/98/oj

European Commission. (1999). *Council Directive 1999/31/EC of 26 April 1999 on the landfill of waste*. https://www.legislation.gov.uk/eudr/1999/31/contents#

European Commission. (2008). *Directive 2008/98/EC of the European Parliament and of the Council of 19 November 2008 on waste and repealing certain Directives*. https://eur-lex.europa.eu/eli/dir/2008/98/oj

IKEA. (2020). *Transforming into a circular business*, https://www.ikea.com/global/en/our-business/people-planet/a-world-without-waste/

Scharff, C. (2018). *The EU circular economy package and the circular economy coalition for Europe*. https://www.cec4europe.eu/wp-content/uploads/2022/01/Scharff-The-EU-Circular-Economy-Package-and-CEC4Europe-_2018.pdf

WasteSMART/EEA. (2023). *Waste and recycling*, https://www.eea.europa.eu/en/topics/in-depth/waste-and-recycling?activeTab=fa515f0c-9ab0-493c-b4cd-58a32dfaae0a

Watkins, E., & Meysner, A. (2022). *European Circular Economy policy landscape overview', Report, Institute for European Environmental Policy.*

Exploring the Ambitions and Needs of Small and Medium-Sized Enterprises in the Transition to the CE in Vietnam

Thanh Cong Nguyen and Timber Haaker

Abstract The survey results of 69 small and medium-sized enterprises (SMEs) located in the northern, central, and southern regions of Vietnam are reported in this study. The surveyed enterprises show their ambitions of implementing the circular economy (CE) models in the future. The key motivators for circular business model adoption from the SMEs' perspectives fall into economic, financial, and market categories. The key barriers to adopting circular business models relate to technological and cultural factors. SMEs in Vietnam still lack experience and knowledge of circular business models. The survey results show a strong need for supporting SMEs in Vietnam to gain a better understanding of the CE approach. The findings from our survey can support a more effective CE adoption and implementation in developing countries.

Keywords Circular economy transition · Small and medium-sized enterprises · Circular business model · Vietnam

1 Introduction

The projected increase of the world population to approximately nine billion is bound to exert pressure on resources to fulfil future demand. A higher population entails more consumption of durable and nondurable goods and services. This leads to challenges of environmental pollution and waste generation, climate change, the increase of greenhouse gases, and the exhausting exploitation of natural resources. To tackle these issues, countries around the world, including Vietnam, are gradually transitioning to green and sustainable economic models.

T. C. Nguyen (✉)
National Economics University, Hanoi, Vietnam
e-mail: thanhnc@neu.edu.vn

T. Haaker
Saxion University of Applied Sciences, Enschede, The Netherlands
e-mail: t.i.haaker@saxion.nl

A. Heshmati, P. K. Nam (eds.), *Circular Economy and its Implementations in Southeast Asia*, Frontiers in South and Southeast Asian Development Research,
https://doi.org/10.1007/978-981-96-0539-2_11

Promoting sustainable land use, protecting ecosystem recovery, and increasing the use of renewable energy sources are all directly related to CE principles that help conserve natural capital. The Ellen MacArthur Foundation (EMF) suggests that the circular economy (CE) involves reducing the consumption of finite resources and designing products to gradually remove waste from the system. The CE model will help increase economic, natural, and social capital simultaneously. This model represents a systemic shift to build long-term resilience, create business and economic opportunities, and deliver environmental and social benefits (EMF, 2012). The CE approach requires sustainable actions at every stage including extraction, production, distribution, use, and recycling of products, components, and materials.

To achieve the requirements of a CE, the behavior of both producers and consumers needs to change. Government should act as a facilitator for stakeholders by taking on roles related to governance, advocacy, and policymaking. The combination of behavioral change and effective policies will help to encourage consumers and manufacturers to move toward more environmentally friendly products. In addition, policies should also aim to encourage consumers to recycle and reuse products as much as possible, thereby motivating recyclers to do their part and allowing each person to become part of the CE system.

There are significant differences in the adoption and implementation of the CE concept between developed and developing countries. CE first gained attraction and is more commonly applied by developed countries, particularly in the European region with most studies, initiatives, achievements, and collaboration among the members of the European Union. The UK, Germany, Italy, and France have had the highest average circular investments throughout 2006–2016 out of 28 countries in the EU (Marino & Pariso, 2020). Studies on the framework, implementation actions, and opportunities for improvement of the CE are undertaken with a concentration in developed countries (Halog & Anieke, 2021). In contrast, little attention has been paid to the CE approach in developing countries where consumers and producers are still attached to the linear consumption model (Abdul-Hamid et al., 2020; Ting et al., 2023; Thanh et al., 2022). However, existing CE conceptualizations are based on formally organized developed markets.

In practice, the Global South has informal low-income contexts, which are rich in necessity-driven value-retaining practices for materials and goods (Korsunova et al., 2022). Nowadays, moving toward a sustainable development path is the goal shared by many countries and organizations. There is an urgent need for developing countries to adopt and implement the CE approach for long-term economic growth, environmental protection, and social inclusiveness. Enterprises play a crucial role in implementing the CE approach. Hence, this chapter focuses on the enterprises' adoption of circular business models and circular practices in Vietnam.

Small and medium-sized enterprises (SMEs) play an important role in Vietnam's economy. While SMEs account for 96% of the total number of businesses, they employ 47% of the labor force and contribute 36% of the national value added (OECD, 2021). Hence, SMEs in Vietnam would be significant partners in the transition, adoption, and operation of circular business models. A better understanding of SMEs' current position toward the CE approach could accelerate the CE transition in Vietnam.

It is challenging to implement the CE approach, especially in the developing country context. Kirchherr et al. (2018) found that the most pressing CE barriers for firms in the European Union are related to cultural and regulatory context, market, and technology. The finding is consistent with the review by Grafström and Aasma (2021) on publications not limited to the European Union. Apart from understanding the barriers, the motivators are equally important to catalyze the adoption of the circular business model. Hence, the enterprises' perception on adoption of circular business models needs to be explored.

To be able to understand the current position as well as the needs of the companies and evaluate the criteria with regard to CE, an online survey was carried out within the "Capacity building program for enterprises to accelerate circular economy (CE) transition in Vietnam" project, supported by the Embassy of the Netherlands in Vietnam and the United Nations Development Programme (UNDP).[1] Utilizing the survey results of 69 SMEs in Vietnam, this chapter provides an empirical assessment of where companies currently stand and what their ambitions and needs are regarding the transition toward the CE.

2 Literature Review

2.1 Overview of Circular Business Models

A business model describes how an organization, or network of organizations, seeks to create value for its customers, for itself, and for other stakeholders (adapted from Bouwman et al., 2008). The revenue model is the part of the business model that describes how revenue is generated. A business model consists of multiple components such as the value proposition, the organizational model, and the revenue model (Jonker et al., 2021).

The CE concept revolves around business and revenue models in which value retention of products, components, and materials is central, with as little negative impact as possible (Jonker et al., 2021; Lüdeke-Freund et al., 2018). Business models offer a range of (strategic) approaches to form various ways of value retention leading to plenty of value-creation opportunities. Business models provide the basis for transactions between parties (e.g., Business2Consumers, Consumers2Consumers, or Busines2Business). This implies that there is an exchange, a performance, and a counterbalance. In the process of value creation based on transactions, several values are always created simultaneously for and by

[1] For more information, please refer to the UNDP website: https://www.undp.org/vietnam/press-releases/netherlands-and-undp-cooperate-accelerate-circular-economy-transition-viet-nam (accessed on 03/03/2024).

the parties involved. This is commonly referred to as multiple value creation. It should be noted that values are subjective and context (place and time) and person dependent.

The coming decades will be marked by fundamental social and economic changes toward sustainability. Enterprises can make fundamental contributions to these changes. This means that in their value proposition, in their own organization, and in the value chain, they contribute together with partners to creating sustainability and circularity. This ambition has major consequences for current business models, which are based on the principles of a linear economy. Enterprises must consider making changes toward circular business models.

In recent years, the CE has attracted growing political and economic attention (Ekins et al., 2020). According to Kirchherr et al. (2017), "a CE describes an economic system that is based on business models which replace the 'end-of-life' concept with reducing, alternatively reusing, recycling and recovering materials in production/distribution and consumption processes, [thus operating at the micro level (products, companies, consumers), meso level (eco-industrial parks) and macro level (city, region, nation and beyond)], with the aim to accomplish sustainable development, which implies creating environmental quality, economic prosperity and social equity, to the benefit of current and future generations." This is a lengthy but a complete definition of the CE in the sense that all relevant dimensions are included and that it has the "business model" as a central concept for organizing circularity. The definition of CE from Kirchherr et al. (2017) is useful to understand the challenges of the circular transition in both the Global South and Global North as it is focused on business models, on the creation of multiple values, and on operations on both company and more systemic levels.

At the heart of the CE is organizing the value retention of products, components, and materials in loops. The concept of loops should not be taken too literally. Materials by no means always return to the same form or product but transform during the life cycle into other forms and values and in various successive loops. A powerful visual image of the different forms of lifetime extension in loops concerns the so-called R-strategies, i.e., concrete ways of either closing, slowing, or even narrowing material loops (Fig. 1).

Improvement of material efficiency with the application of the CE approach in the supply chain potentially increases long-term income generation (Franklin-Johnson et al., 2016). The circular business model has the capacity to maintain input resources independent of market volatility and to provide value for businesses even after production has finished (Veleva et al., 2017). In this case, the main benefits come from lowering the costs of obtaining materials and boosting income from products or materials that are at the end of their life cycle (Geissdoerfer et al., 2020).

10 Refuse	Preventing the use of 'virgin' (processed) raw materials
9 Reduce	Reduce the use of raw and processed materials
8 Redesign	(Re)design a product or component with sustainability and circularity in mind
7 Re-use	Reusing products, components or (processed) raw materials
6 Repair	(Timely) maintenance and repair whether or not combined with redesign and digitalization
5 Refurbish	Refurbish products and components such that they are 'like new'
4 Remanufacture	Create new products or components from previously made products or components
3 Repurpose	Reuse products or components but with a different purpose/function
2 Recycle	Processing of products and components to raw material and reuse
1 Recover	Energy recovery from materials

Fig. 1 Overview of R-strategies (adapted from Jonker et al. 2021)

2.2 *Motivators and Barriers to Adoption of Circular Business Models*

Companies find several reasons to work toward more circular practices. These range from a sense of responsibility for future generations, requirements from customers and investors to increasing legislation and regulation (Jonker et al., 2021). As more organizations across different countries begin to understand the importance of the CE, more institutional frameworks are designed to promote the adoption of circular business models. Then, the laws and regulations have created a motivating environment for enterprises to adopt the CE approach. Jesus and Mendonça (2018) classify motivators for CE adoption into a few categories, namely, economic, financial, market, institutional, and regulatory factors. While The economic factors include availability of raw materials, access to appropriate technology, the financial motivators mean access to adequate funding and other financial benefits. The market represents consumers' inclination toward greener products and services, leading to both products and services being produced and priced accordingly in a sustainable manner. The Institutional and regulatory motivators indicate enactments and enforcement of laws and regulations for cleaner and sustainable manufacturing.

Motivators are also classified into internal and external drivers. Internal factors are related to pressure from shareholders, support from the management, skilled labor, connectivity among companies, and technology processes (Jabbour et al., 2020). Technology is a key factor for successful adoption of a circular business model—hence, this business model offers incentives for technology development

(Hina et al., 2022). Nowadays, various types of technology, including blockchain, the Internet of Things (IoT), radio-frequency identification (RFID), machine learning, big data, and artificial intelligence, are the enablers of circular supply chain management (Tura et al., 2019). The adoption of digitization and intelligent robotics, which improve efficiency in waste sorting and treatment systems, is prevalent nowadays (Kirchherr et al., 2018).

For external factors, the key stakeholders are mainly customers, suppliers, investors, competitors, and government policymakers. Enterprises adopting a circular business model can enhance their reputation in the eyes of consumers and develop their value networks with partners (Shao et al., 2020). As aforementioned, the material loops of the circular business model can operate at the meso level (eco-industrial parks) where enterprises create connections in utilizing scrap materials in the industrial zone (Kirchherr et al., 2017). In that case, enterprises will actively interact with suppliers in designing a circular supply chain network (Lewandowski, 2016).

In addition, innovation is an important factor to successfully implement circular business models. A circular business model needs new ideas, renewable energy systems, and infrastructure to improve efficiency of the traditional systems. Innovation also can reduce the burden of incurred costs and enhance the possibility of interacting enterprises in creating industrial symbiosis and a value network for circulating materials of enterprises.

There is also concern about the resource scarcity which was the main motivational factor for developed nations with limited resources, particularly within the European region. Tura et al. (2019) considered resource scarcity as one of the main factors encouraging firms to adopt the CE concepts to recirculate scrap resources into a production and consumption cycle. The driver of resource scarcity is shared by countries and organizations around the world, as can be seen by the commitment of United Nations' SDGs to responsible production and consumption effort.

The implementation and adoption of the circular business model also face various obstacles. Kirchherr et al. (2018) discussed the barriers to CE adoption in a few categories, namely, cultural, regulatory, market, and technology. The cultural barriers can be considered as "lacking consumer interest and awareness," "hesitant company culture," and "operating in a linear economy." Currently, as the concept of CE is gaining more and more attention, many countries have set up their institutional frameworks for promoting the circular business model, but these efforts appear to be ineffective owing to the "culture of hesitancy" among businesses. The hesitance in adopting the circular business model may stem from corporate executives who lack entrepreneurial spirit or understanding of the circular business model. Even though enterprises' executives support corporate social responsibility by adopting the circular business model, it is difficult to achieve a balance of social, economic, and environmental dimensions if they lack a clear vision or knowledge of this model. Regulatory factors come in the form of inconsistent policy and guidelines from local or global authorities resulting in a lack of consensus as to what constitutes the adoption of a CE. The market factor involves initial high investment costs and difficulty to access funds to support the circular business models. Technological barrier denotes the lack of proven track records of successful large-scale projects to serve as a benchmark which helps explain why funding is difficult to secure since data on the benefits of circular business models are limited in nature.

Barriers to a CE can also be classified into a few groups which are financial, structural, operational, attitudinal, and technological barriers (Jabbour et al., 2020; Ritzén & Sandström, 2017). Out of the five barriers, three came from cultural barriers in a study by Kirchherr et al. (2018). This indicates the existence of differences in classifying CE barriers between academic researchers and business practitioners.

The circular business model is mostly applied by large corporations, with no sign of small businesses (Lewandowski, 2016). Although governments throughout the globe have encouraged small and medium-sized firms to embrace the circular business model, many businesses still face the difficulty of gaining clear ideas of the model, mobilizing financial resources, and incurring excessive conversion expenses. There may be a case for technology import as a means for small and medium-sized businesses to migrate to a circular business model (Geissdoerfer et al., 2020). On the other hand, this method may provide a long-term problem if it requires foreign professionals to operate and resolve technical faults.

2.3 *The Transition to the CE in Vietnam*

In Vietnam, the transition toward the CE has been paid attention in recent years. At national level, the concept of CE has been specifically stipulated in Article 142 of the Environmental Protection Law 2020 and in Article 139 of Decree No. 08/2022/ND-CP on Detailed Regulations for Articles of Environmental Protection Law. Along with that, in 2022, the government approved the CE Development Scheme in Vietnam (Decision 687/2022/QD-TTg). Currently, the Ministry of Natural Resources and Environment is assigned to preside over and coordinate with other ministries, ministerial-level agencies, and provincial-level People's Committees to develop and submit to the National Action Plan for CE Implementation by the end of 2023.

At micro level, enterprises have started considering the CE approach in their business models due to regulations and market requirements. However, the level of application of CE in Vietnamese enterprises is relatively low (CIEM, 2022). CIEM (2022) reported that the percentage of enterprises applying CE in five forms including "functional sales," "cradle to cradle," "green supply chain management," "industrial symbiosis," and "take-back mechanism" at a good level only account for 3–6%. Meanwhile, the percentage of businesses applying circular business models including "repair and maintenance," "reuse and redistribution," "refurbishment and remanufacturing," "recycling," "cascading and repurposing," and "organic feedstocks" at good levels are ranging from 3.3 to 5.5% (CIEM, 2022). In addition, the percentage of enterprises that have never applied any form of business model innovation and/or circular business model is 37.6%. Thus, the level of application of circular business models is still low. Most Vietnamese enterprises have not applied or even learned about CE, especially SMEs. Only about 3–6% of enterprises surveyed said that they have applied innovative business models or only applied one form of circular business model (CIEM, 2022).

3 Research Design

The CE concept is popular among developed countries contributing to sustainable production, efficient resource utilization, a new economic model, and higher skilled job creation. With the gap in developing countries within the Southeast Asian region, this research's main goal is to understand the current position as well as the needs of SMEs in Vietnam with regard to the adoption of circular business models. Our survey aims to explore where companies currently stand and what their ambitions and needs are regarding the transition toward circular business models.

A descriptive research design was suitable for this study as the research topic frequently uses data collection methods such as questionnaire survey that involves "asking respondents structured questions about what they think, feel and do" (Hair et al., 2021, p. 109). Data was collected from primary sources through a structured questionnaire survey. The questionnaire was designed based on an adaptation of the *Quick Scan Circular Business Models* questionnaire from Jonker et al. (2021).

It consists of questions divided into two main sections: (1) general information about the enterprises and (2) perspectives of the enterprise on CE. Our exploration of enterprises' perspectives focuses on their motivations, ambitions, experiences, and choices of circular business models. Perception questions were designed according to the five-point Likert scale, for example, how strongly respondents agree or disagree with the statements (1 = strongly disagree and 5 = strongly agree). Finally, respondents were asked what support they need to enhance their capacity to adopting circular business models.

The survey was implemented within the "Capacity building program for enterprises to accelerate circular economy (CE) transition in Vietnam" project with the support of Hue Innovation Hub, the Institute for CE Development (ICED), and the Environmental Economics and Policy Institute (EEPI). The online questionnaire was delivered using a Google Form in June 2022.

4 Results

The survey collected responses from 69 small and medium-sized enterprises located in the northern, central, and southern regions of Vietnam. The main results are presented in the following sections.

4.1 Characteristics of Enterprises in the Survey Sample

Some basic information about the surveyed enterprises is summarized in this section. The average number of regular employees in each enterprise is 74 people. As per the law to support small and medium-sized enterprises (SMEs) in Vietnam, SMEs are defined as micro, small, and medium-sized enterprises having no more than 200 employees registered with the state social insurance scheme in a year. Most of our sample (92% of the surveyed enterprises) are SMEs, with 45% of the surveyed enterprises categorized as microenterprises with about 10 employees, 44% are small enterprises with less than 100 employees, and 3% are medium-sized businesses with about 200 employees.

Most enterprises participating in the survey (93%) are private enterprises, only 4% of the surveyed enterprises are enterprises with state capital, and 3% are foreign invested companies (Fig. 2). The industries of enterprises participating in the survey are very diverse, most of which are manufacturing enterprises, agricultural and food-processing businesses, and waste treatment companies.

4.2 Enterprises' Opinions on Circular Business Models

Table 1 presents enterprises' level of agreement with the statements related to their motivations, ambitions, experiences, and choices of circular business models. The level of agreement was designed according to the five-point Likert scale, which is 1 = strongly disagree and 5 = strongly agree.

According to Jesus and Mendonça (2018), motivators for CE adoption can be classified into a few categories, namely, economic, financial, market, and institutional/regulatory factors. In our survey, enterprises indicate that their main reasons for considering circular business models are (i) to reduce the consumption of raw materials, (ii) to enhance their market position, (iii) to achieve their ambition, (iv) to avoid problems with resource scarcity in the future, and (v) to obtain financial

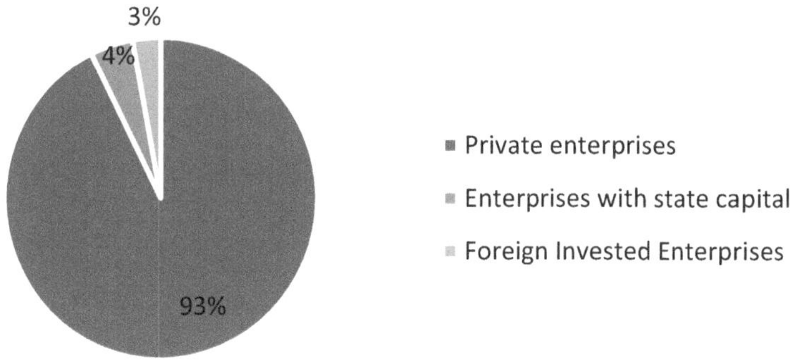

Fig. 2 Types of surveyed enterprises

Table 1 enterprises' motivations, ambitions, experiences, and choices of circular business models

Statements		Mean	Standard deviation
Why is circular business important to your company? ***(Motivation)***			
1	To reduce the consumption of raw materials	4.13	0.90
2	Working on sustainability and circularity has a positive impact on our market position	4.10	0.78
3	Our ambition is to be sustainable and circular	4.03	0.90
4	To avoid problems with resource scarcity in the future	3.97	0.99
5	It is financially attractive due to subsidies	3.70	0.98
6	To avoid legal troubles and environmental penalties	3.62	1.01
7	To avoid conflicts with surrounding communities	3.39	1.01
8	We will be forced by legislation soon	3.36	0.99
9	Our customers will increasingly expect it of us	3.10	1.08
10	To be able to recruit workers	3.04	1.00
How would you like to develop further in the field of sustainable and circular business in the coming years? ***(Ambitions)***			
1	In about 10 years, we want to be a circular and sustainable business	4.06	0.88
2	We are developing environmentally friendly and circular products	4.00	1.04
3	We are gradually adapting to laws and regulations	3.88	0.79
4	We actively identify circular challenges and opportunities	3.88	0.96
5	We have clear plans for the coming years	3.67	0.95
What have you done so far in terms of circular business? ***(Experiences)***			
1	We strive to reduce the consumption of raw materials, thereby reducing waste at the same time	3.91	0.97
2	For us, circular business is not a choice but a strategic guiding principle	3.90	0.96
3	We started to develop our value chain in a circular way, in cooperation with parties (suppliers or customers) in the chain	3.67	0.98
4	Circular business is already familiar to us	3.49	0.97
5	We measure aspects of circularity in our company	3.33	1.02
Which of the following strategies of circular business model are you already implementing or planning? ***(Choice of strategies)***			
1	Reuse: Reusing products, components, or (processed) raw materials	3.91	0.94
2	Repair: (Timely) maintenance and repair whether or not combined with redesign and digitalization	3.84	1.04
3	Reduce: Reducing the use of (processed) raw materials	3.81	0.86
4	Rethink: (Re)designing a product or component with sustainability and circularity in mind	3.77	0.97
5	Remanufacture: Create new products or components from previously made products or components	3.75	1.03
6	Recycle: Processing of products and components to raw material and reuse	3.72	1.02
7	Repurpose: Reuse products or components but with a different purpose/function	3.71	1.03
8	Apply online sharing platforms	3.68	1.03

(continued)

Table 1 (continued)

Statements		Mean	Standard deviation
9	Recover: Energy recovery from materials	3.65	1.01
10	Connect with partners to share data, share production capacity	3.65	1.00
11	Refurbish: Products and components refurbished "after which they are like new"	3.59	1.04
12	Applying product as a service (e.g., lease instead of sale)	3.59	1.04
13	Refuse: Preventing the use of "virgin" (processed) raw materials	3.55	1.04

advantages, such as subsidies. It appears that the key motivators for circular business model adoption from the SMEs' perspectives fall into economic, financial, and market categories.

Exploring the ambition of implementing circular business in the future, the survey results indicate that the top 3 aims of enterprises participating in the survey are (1) to be recognized as a circular and sustainable business in the next 10 years, (2) to focus on developing environmentally friendly and circular products, and (3) to gradually adapt to the requirements of legal regulations.

In relation to enterprises' experiences, the results show that the surveyed enterprises are striving to reduce the consumption of raw materials and have started to develop their value chain in a circular way, in cooperation with partners (suppliers or customers) in the chain. For the surveyed SMEs, circular business also has been identified as a strategic guiding principle.

The survey results also provide information on circular business models being implemented or planned to adopt. The top selected business models include reuse, repair, reduce, rethink, and remanufacture. However, the scoring results are usually less than 4 (that is considered as "agree"), indicating that the circular business models have not been widely adopted in the current operation of the surveyed enterprises.

To examine consistency in enterprises' responses, the correlations between enterprises' choices of circular business models and their motivations, ambitions, and experiences were estimated for the top 3 responses with the highest mean scores of agreements (Table 2). The correlations are all statistically significant. Hence, it appears that enterprises' responses on choices of circular business models are consistent with their motivations, ambitions, and experiences.

To gain better understanding of enterprises' needs, enterprises joining our survey were requested to name three important factors that their business needs to address in order to achieve their goal of adopting circular business models. The challenging factors reported by enterprises in our survey were classified into four groups of barriers, namely, cultural, regulatory, market, and technology, as suggested by Kirchherr et al. (2018). Figure 3 shows a ranking of barriers based on how frequently surveyed enterprises reported the barriers. Enterprises' responses in our survey indicate that the key barriers to the adoption of circular business models relate to technological and cultural factors.

Table 2 Correlations between the top 3 responses on choices of circular business models and motivations, ambitions, experiences

Variables	Choice of reuse model	Choice of repair model	Choice of reduce model
Motivation1: To reduce the consumption of raw materials	0.291*	0.392**	0.295*
Motivation2: Positive impact on our market position	0.380**	0.529**	0.472**
Motivation3: Ambition is to be sustainable and circular	0.356**	0.335**	0.333**
Ambition1: In about 10 years, we want to be a circular and sustainable business	0.619**	0.507**	0.601**
Ambition2: Developing environmentally friendly and circular products	0.558**	0.549**	0.523**
Ambition3: Gradually adapting to laws and regulations	0.441**	0.346**	0.396**
Experience1: To strive to reduce the consumption of raw materials, thereby reducing waste at the same time	0.577**	0.565**	0.492**
Experience2: Circular business is a strategic guiding principle	0.499**	0.393**	0.386**
Experience3: To develop our value chain in a circular way	0.561**	0.490**	0.534**
Number of observations	69	69	69

Notes: **Correlation is significant at the 0.01 level (two-tailed). *Correlation is significant at the 0.05 level (two-tailed)

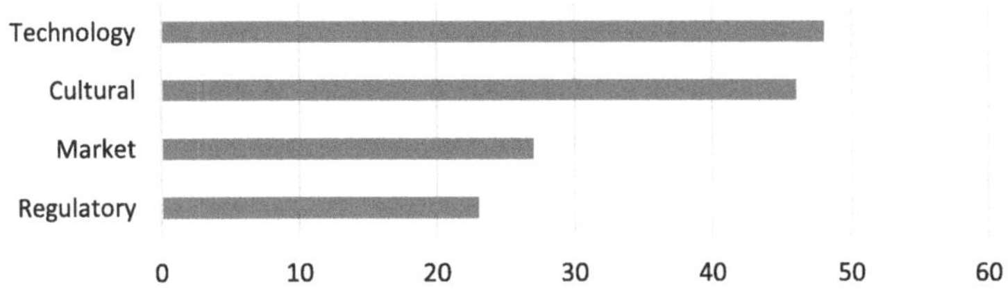

Fig. 3 Ranking of barriers

There is a strong need for a better understanding of the CE approach from SMEs in Vietnam. As shown in Table 3, some examples of the enterprises' needs are as follows:

- Better understanding of the CE in order to be able to make decisions to implement a circular business model
- Finding the right partners to jointly develop circular products
- Estimating the benefits of enterprise's CE practices
- Learning about market expansion opportunities related to the implementation of the CE approach
- Learning about the criteria for evaluating the CE in enterprises' industry sector

Table 3 Enterprises' needs for enhancing their capacity of adopting circular business models

Statements		Mean	Standard deviation
To enhance capacity of adopting circular business models, what do you need to support?			
1	The need to better understand the CE in order to be able to make decisions to implement a CE	4.16	0.91
2	Finding the right partners to jointly develop cyclical products	4.13	0.88
3	Estimating the benefits of enterprise's CE practices	4.12	0.88
4	Learn about market expansion opportunities, and build an image related to the implementation of the CE	4.10	0.98
5	Learn about the criteria for evaluating the CE in your industry	4.09	0.97
6	Looking for opportunities to get support in training employees to implement CE	4.07	0.89
7	Determine the right CE strategy for your business	4.06	0.99
8	Improve the existing CE implementation strategy so that it can be improved and upgraded	4.06	0.89
9	Update legal regulations related to CE	4.04	0.88
10	Looking for partners to supply raw materials to meet the enterprise's CE implementation strategy	4.04	0.92
11	Looking for suitable partners to exchange and share materials	4.04	0.91
12	Seeking technology-supported opportunities to implement CE	4.04	0.95
13	Looking for financial support to implement CE	4.01	0.94

5 Conclusions

By maintaining resources in circulation for as long as feasible, a circular business model contributes to the economic growth of firms, society, and the environment (Ranta et al., 2018). It is one of the most important foundations that determine the success of the CE.

This chapter presents the survey results of 69 SMEs located in the northern, central, and southern regions of Vietnam. The surveyed enterprises show their ambitions of implementing the CE models in the future.

The key motivators for circular business model adoption from the SMEs' perspectives are found to fall into economic, financial, and market categories. The key barriers to the adoption of circular business models relate to technological and cultural factors.

Building an environmentally friendly and sustainable image along with the uncertainty elements in CE adoption is relevant to many firms in developing countries. SMEs in Vietnam still lack experience and knowledge of circular business models. The results show a strong need for supporting SMEs in Vietnam to gain a better understanding of the CE approach. The findings from this chapter can be an indicator and tool for policymaking in developing countries for a more effective CE adoption and implementation.

References

Abdul-Hamid, A.-Q., Ali, M. H., Tseng, M.-L., Lan, S., & Kumar, M. (2020). Impeding challenges on industry 4.0 in circular economy: Palm oil industry in Malaysia. *Computers & Operations Research, 123*, 105052.

Bouwman, H., de Vos, H., & Haaker, T. (2008). Mobile service innovation and business models. Springer-Verlag Berlin Heidelberg. https://doi.org/10.1007/978-3-540-79238-3.

CIEM. (2022). *The circular business model: International experience and application in Vietnam (in Vietnamese).* Central Institute for Economic Management (CIEM).

Ekins, P., Domenech Aparisi, T., Drummond, P., Bleischwitz, R., Hughes, N., & Lotti, L. (2020). *The circular economy: What, why, how and where.* The OECD Centre for Entrepreneurship.

EMF. (2012). *Towards the circular economy: An economic and business rationale for an accelerated transition.* The Ellen MacArthur Foundation (EMF).

Franklin-Johnson, E., Figge, F., & Canning, L. (2016). Resource duration as a managerial indicator for circular economy performance. *Journal of Cleaner Production, 133*, 589–598.

Geissdoerfer, M., Pieroni, M. P. P., Pigosso, D. C. A., & Soufani, K. (2020). Circular business models: A review. *Journal of Cleaner Production, 277*, 123741.

Grafström, J., & Aasma, S. (2021). Breaking circular economy barriers. *Journal of Cleaner Production, 292*, 126002.

Hair, J. F., Celsi, W. C., Ortinau, D. J., Bush, R. P. (2021). *Essentials of marketing research, 4th edition.* New York, NY: McGraw-Hill Higher Education.

Halog, A., & Anieke, S. (2021). A review of circular economy studies in developed countries and its potential adoption in developing countries. *Circular Economy and Sustainability, 1*(1), 209–230.

Hina, M., Chauhan, C., Kaur, P., Kraus, S., & Dhir, A. (2022). Drivers and barriers of circular economy business models: Where we are now, and where we are heading. *Journal of Cleaner Production, 333.*

Jabbour, C. C., Jose, S. S., de Sousa Jabbour, A. B. L., Jugend, D., Fiorini, P. D. C., Latan, H., & Izeppi, W. C. (2020). Stakeholders, innovative business models for the circular economy and sustainable performance of firms in an emerging economy facing institutional voids. *Journal of Environmental Management, 264*, 110416.

Jesus, A., & Mendonça, S. (2018). Lost in transition? Drivers and barriers in the eco-innovation road to the circular economy. *Ecological Economics, 145*, 75–89.

Jonker, J., Faber, N., & Haaker, T. (2021). *Quick scan circular business models, inspiration for organising value retention in loops.* Ministry of Economic Affairs and Climate Policy.

Kirchherr, J., Reike, D., & Hekkert, M. (2017). Conceptualizing the circular economy: An analysis of 114 definitions. *Resources, Conservation and Recycling, 127*, 221–232.

Kirchherr, J., Piscicelli, L., Bour, R., Kostense-Smit, E., Muller, J., Huibrechtse-Truijens, A., & Hekkert, M. (2018). Barriers to the circular economy: Evidence from the European Union (EU). *Ecological Economics, 150*, 264–272.

Korsunova, A., Halme, M., Kourula, A., Levänen, J., & Lima-Toivanen, M. (2022). Necessity-driven circular economy in low-income contexts: How informal sector practices retain value for circularity. *Global Environmental Change, 76*, 102573.

Lewandowski, M. (2016). Designing the business models for circular economy—Towards the conceptual framework. *Sustainability.*

Lüdeke-Freund, F., Carroux, S., Joyce, A., Massa, L., & Breuer, H. (2018). The sustainable business model pattern taxonomy—45 patterns to support sustainability-oriented business model innovation. *Sustainable Production and Consumption, 15*, 145–162.

Marino, A., & Pariso, P. (2020). Comparing European countries' performances in the transition towards the circular economy. *Science of the Total Environment, 729*, 138142.

OECD. (2021). *SME and entrepreneurship policy in Viet Nam.* OECD Publishing.

Ranta, V., Aarikka-Stenroos, L., & Mäkinen, S. J. (2018). Creating value in the circular economy: A structured multiple-case analysis of business models. *Journal of Cleaner Production, 201*, 988–1000.
Ritzén, S., & Sandström, G. Ö. (2017). Barriers to the circular economy—Integration of perspectives and domains. *Procedia CIRP, 64*, 7–12.
Shao, J., Huang, S., Lemus-Aguilar, I., & Ünal, E. (2020). Circular business models generation for automobile remanufacturing industry in China. *Journal of Manufacturing Technology Management, 31*(3), 542–571.
Thanh, T. T., Tu, A. N., Phuong, T., & Thu, H. N. (2022). Advancing the circular business models in developing countries: Lessons from China. *Green and Low-Carbon Economy.*
Ting, L. S., Zailani, S., Sidek, N. Z. M., & Shaharudin, M. R. (2023). Motivators and barriers of circular economy business model adoption and its impact on sustainable production in Malaysia. *Environment, Development and Sustainability.*
Tura, N., Hanski, J., Ahola, T., Ståhle, M., Piiparinen, S., & Valkokari, P. (2019). Unlocking circular business: A framework of barriers and drivers. *Journal of Cleaner Production, 212*, 90–98.
Veleva, V., Bodkin, G., & Todorova, S. (2017). The need for better measurement and employee engagement to advance a circular economy: Lessons from Biogen's "zero waste" journey. *Journal of Cleaner Production, 154*, 517–529.

The manufacturer's authorised representative in the EU is Springer Nature Customer Service Centre GmbH, Europaplatz 3, 69115 Heidelberg, Germany. If you have any concerns regarding our products, please contact ProductSafety@springernature.com

Printed and bound by CPI Group (UK) Ltd, Croydon, CR0 4YY
01/10/2026
02225938-0001